*A Gift for*

_____

*Presented by*

_____

$$E=MC^2$$

"There is something fascinating about science. One gets such wholesale returns of conjecture out of such a trifling investment of fact."

—Mark Twain

# $E=MC^2$

## Simple Physics
Why Balloons Rise, Apples Fall,
and Golf Balls Go Awry

JEFF STEWART

The Reader's Digest Association, Inc.
New York, NY/Montreal

*To Mom and Dad and Rosemary and Terry,*
*thank you for helping create the space-time I needed to write.*

*Thank you Herbie, Iola, Rufus and Oscar for*
*making sure I took regular breaks.*

*Thank you Lindsay for making the whole thing happen.*

A READER'S DIGEST BOOK

Copyright © 2010 Michael O'Mara Books Limited

All rights reserved. Unauthorized reproduction, in any manner, is prohibited.

Reader's Digest is a registered trademark of The Reader's Digest Association, Inc.

First published in Great Britain by Michael O'Mara Books Limited
9 Lion Yard, Tremadoc Road, London, SW4 7NQ

Illustrations by David Woodroffe
Designed by Ed Pickford

FOR READER'S DIGEST
U.S. Project Editor: Siobhan Sullivan
Copy Editor: Barbara McIntosh Webb
Canadian Project Manager: Pamela Johnson
Canadian Project Editor: Jesse Corbeil
Indexer: Andrea Chesman
Project Production Coordinator: Nick Anderson
Senior Art Director: George McKeon
Executive Editor, Trade Publishing: Dolores York
Manufacturing Manager: Elizabeth Dinda
Associate Publisher, Trade Publishing: Rosanne McManus
President and Publisher, Trade Publishing: Harold Clarke

Library of Congress Cataloging-in-Publication Data:

Stewart, Jeff, 1970-
E=MC2 : simple physics : why balloons rise, apples fall &
golf balls go awry / Jeff Stewart.
   p. cm.
ISBN 978-1-60652-167-0
1. Physics--Popular works. I. Title. II. Title: Simply physics.
QC24.5.S746 2010
530--dc22

                              2010019429

We are committed to both the quality of our products and the service
we provide to our customers. We value your comments, so please feel
free to contact us.

The Reader's Digest Association, Inc.
Adult Trade Publishing
44 S. Broadway
White Plains, NY 10601

For more Reader's Digest products and information, visit our website:
www.rd.com (in the United States)
www.readersdigest.ca (in Canada)

Printed in the United States of America

3 5 7 9 10 8 6 4

# CONTENTS

1 A Brief History of Physics     7

2 Forces     23

3 Energy and Power     41

4 Momentum     57

5 Heat and Matter     75

6 Waves     89

7 Electricity     105

8 Relativity     119

9 Quantum Physics     137

10 The Universe     155

Index     171

Metric Conversion Table     175

# 1

# A Brief History of Physics

What exactly *is* physics?

Well, in a way, it's everything.

Physics aims to tell us about a big bang that created the universe long ago, to explain how people got here from there (and why we won't be going anywhere else in a hurry), and to show how and why everything around us works as it does.

It tells us how the first bits of matter appeared, how the first stars were born, and how, over billions of years, the universe came to be the vast and violent place we know, with our planet an insignificant speck on the edge of one galaxy in 125 billion.

It explains almost everything that happens in the world around us: energy and movement, sound and light, electricity and matter. And its laws form the basis of chemistry and biology.

Physics also suggests exciting new ideas. For example, it says that time travel may be possible. Unfortunately, it also says that we're probably too big to try it.

In short, modern physics gives us a fascinating, awe-inspiring, and sometimes downright weird view of the universe and our place in it.

Falling apples, rising balloons, and errant golf balls are just the start.

## It's the law

Physicists—a mixture of mathematical thinkers and more practical types who enjoy doing experiments, such as smashing tiny bits of stuff into even tinier bits—think that they can explain all this because everything happens according to the laws of nature.

These laws show that, if this thing happens, then so will that. If I hang a weight (me, for example) from a spring, the amount it stretches will be proportional to my weight: double the weight, double the stretch. (This particular law is known as Hooke's Law because it was discovered by the 17th-century British physicist Robert Hooke.)

The laws of physics are useful to us because physics is a practical science. What we've learned has helped us build everything from bathroom scales (that spring again) to a billion shiny gadgets, from the bulbs that light up our cities to the airplanes we take to fly between those cities.

Of course, it has also brought us enough nuclear warheads to blow all of this—and us, and life as we know it—to pieces.

## How we found physics

People have always tried to explain and predict the world around them. It seems to be an essential part of what makes us human. But it wasn't until we got past blaming everything from lightning to earthquakes on a bunch of irritable gods that we started getting anything useful out of our explanations.

So that's what physics is. But a quick run through 2,500 years of scientific progress will give us a better idea of how it got us here. And, hopefully, prove to anyone still frightened by the subject that it doesn't bite—even if it does bang.

## It's all Greek

Almost 2,500 years ago, the ancient Greeks did a lot of thinking about science. Besides running around naked at the first Olympic games, writing tall tales about gods, cunning heroes and many-headed monsters, and building wonderful temples, they came up with plenty of interesting theories.

For example, Thales supposed that all the earth floated on water, so that earthquakes were caused by waves. Aristotle, whose *Physica* is the first work on physics to use the word in the title, believed that everything in the world was made up of earth, air, fire, and water, with the heavens made of a divine substance called ether. Smoke, he thought, rose, because it was mainly made up of air, and air always tended to be above earth.

These were nice simple theories, but the Greeks generally argued that an object does something because it's the kind of object that does that kind of thing. This gets us nowhere: it's a circular argument; a good example of what modern physics isn't. (Today, we try to explain things in terms of other things, which is a nice way of saying we usually like to blame someone else.)

Strictly speaking, the Greeks said that, for example, all circles we see are somehow shadows of the "Form" of roundness. Forms were supposed to be divine, perfect and not of this world, so that the Form of roundness set the perfect example for all other round things. But, it seems, there's no getting around the fact that an object's roundness is still explained by the fact that it is, well, round. Which is, as we noted, circular.

## Mind games

Part of the problem was that the ancient Greeks tended to think that you should be able to figure out what happens in the world through the power of thought alone. The world around us was imperfect, so, they thought, there wasn't much point looking at it too closely and expecting it to behave in a regular way.

And even when they did observe what was going on around them, they made some strange mistakes. For example, Aristotle made detailed studies of plants and animals, but thought that as a rule, men have more teeth than women.

## Predicting the planets

A few hundred years later, another ancient Greek called Ptolemy (ignore that first "p" when you say his name, or you may be accused of spitting) came up with a fairly accurate mathematical system for predicting the position of the stars and planets.

This was a big step forward for science, except he also thought that the planets (and the sun) revolved around the Earth. This meant that to make his numbers work, he had to predict that the moon would sometimes come twice as close to the Earth as at other times. As a result, he thought that we should regularly see the moon appear to double in size.

Of course the moon never grew, but in the Western world, Greek ideas held sway in physics for 1,500 years. Partly this was because the church supported them: Greek thinking, like Christianity, put man at the center of the universe, and with the stars all fixed to the inside of a huge sphere that contained the universe, it also left plenty of room outside for heaven and hell.

As Greek civilization declined, many of its ideas and writings were saved by Islamic scholars, who refined them and slowly increased the importance of math and observation, particularly in the study of light, the stars and motion. This was a good move because, as it turned out, math and observation turned out to be the key to progress in science.

## The scientific revolution

It may seem obvious now, but this new way of thinking revolutionized the way we understand the world. People started to discover that by looking carefully, measuring time and distance and energy, and by putting the numbers together with math and careful thought, we could predict—at least in the laboratory, under carefully controlled conditions—what would happen next and why it would happen. (Actually, it's amazing how far you can get without mathematics, as this book proves. You need math to write your theories neatly and to prove them to other scientists, but usually you can understand the big ideas pretty well without it.)

Astronomy also progressed, with the theory that the Earth revolved around the sun slowly gaining favor, despite the efforts of the Catholic Church. The Bible says that the world is firmly fixed, and this revolutionary new theory was, indeed, revolutionary.

The great Italian physicist Galileo Galilei, who was handy with a telescope and was a dedicated star watcher, offered the first direct evidence for the new theory. For this heresy the church had him placed under house arrest in 1633. He remained there until he died.

Then came Sir Isaac Newton—an English mystic, alchemist, and theologian—a complicated and often very cranky man who was also the greatest mathematician and second greatest physicist ever. In one book, the *Principia Mathematica*, published in 1687, he laid down the law of gravity, as well as three elegant laws of motion that describe how objects move. These were unchallenged for two hundred years and still form the basis of most movement calculations by scientists and engineers.

For example, if you want to know the minimum distance in which you can stop your car when you're doing 80 m.p.h. (130 km/h) on the highway, and you don't want to cause an accident trying it out, all you need is one quick experiment at 20 m.p.h. (32 km/h) on a quiet side street, a copy of Newton's laws of motion, and, for the arithmetically challenged, the ability to multiply and divide on a calculator.

## A clockwork universe

After Newton, it seemed that the universe ran according to a hidden clockwork code, and that it was governed by rules that we can discover and understand. This idea changed the course of human history. It led scientists all over the world to use Newton's mathematical tools to build on his ideas and develop science. And so Newton began the rush of scientific discoveries that continues to this day.

We found out why the Earth goes around the sun and how its heat reaches us. We learned about light, sound, electricity, and energy: how it changes, how it moves, and how it doesn't disappear. We found that everything we see is made up of tiny things called atoms, which are made up of tinier things called electrons and protons and neutrons. And that even then there are tinier things in them.

With this information in hand, engineers began building the modern world.

## At the speed of light

Throughout the 19th century, scientists began to discover tiny flaws in Newton's laws—and the branch of physics known as classical mechanics, which flowed from them. For example, the orbit of the planet Mercury seemed wrong: Its orbit didn't change exactly as the equations said it should. There were also difficulties with handling the speed of light and the way objects radiate heat.

It took the genius of a German patent clerk named Albert Einstein to begin to solve these riddles. In one year, 1905, he wrote four papers that helped physics take huge steps forward—and all while holding down his job at the office. His new ideas included the theory of special relativity (which showed we could never travel faster than the speed of light) and mass-energy equivalence (the famous $E=mc^2$ equation, which showed we could turn objects into energy, and paved the way for nuclear power and atom bombs).

Einstein's theories agreed very closely with Newton's laws under everyday conditions and showed why Mercury went where it did. They also suggested a whole new direction for physics, now called cosmology, which takes it out to work on the entire universe, to tell how stars and even universes might be born.

A hundred years later, Einstein's physics still points the way in cutting-edge research.

## It's neither here nor there

In 1900, German physicist Max Planck solved the radiation of heat problem when he showed that the energy of electromagnetic waves is quantized, so that heat energy is emitted in chunks, which can only get so small and no smaller. In the 1920s his insight led to quantum physics, a whole new branch of the science, which was discovered and fleshed out by a large group of physicists, working at the same time, in different countries, on what often seemed like separate problems. Together, they made quite a mess of the comforting idea of a predictable clockwork universe.

Quantum physics is the science of the very small: of individual light particles, the structure of atoms, and all the things that make them up—not just the electrons and protons that we already knew, but the tinier bits of these bits, such as quarks, all of which have the odd property of being really just waves that are probably here and probably there, though we can never quite be sure.

## New questions

And so we reach the present day. The main thrust of physics today is to put together the theories about the big stuff (Einstein's relativity) and the tiny stuff (quantum physics) into one explanation of the whole universe: how it works, why the laws are as they are, its birth in a big bang, and maybe its death, billions of years in the future.

One version of this goal is called a GUT: a Grand Unified Theory. But whatever it's called, at the moment it's got physics pretty well stumped. The bad news is that it may actually be impossible to create such a theory; or possible, but not possible for human brains.

No one's giving up yet, but these are some of the reasons that some physicists are thinking carefully (or possibly staring out the lab window and daydreaming) about what kind of things their laws are, and how they fit into the universe.

## Wrong again

Of course, physics—and its laws—aren't nearly as straightforward as this history might suggest.

Apart from the risk of nuclear armageddon, the greenhouse effect, and the time we waste deciding which of the many shiny gadgets that physics has spawned we should buy next, there is also the fact that—as we've seen throughout its history, and despite all the trust we put in it to stop our cars—sometimes physics is wrong.

For example, Aristotle thought that heavier things fall faster, which seems to be what we all see: think of feathers and lead weights. But Galileo and Newton proved him wrong.

Then Newton, with the help of falling apples, used the theory of gravity to explain why things fall, and to predict how quickly they gain speed. That was great, but a few hundred years later, Einstein made everything much more complicated.

His theories of relativity say that weight and speed of falling are linked, but in a strange way: the faster things fall, the heavier they get. You would have to be moving many thousands of times faster than a jet aircraft before you'd really notice the weight you put on, but experiments have shown that the theory is true.

So, Einstein's theory is true—and, on the other hand, sometimes it isn't. In quantum physics, where we look at the motion of tiny and often very light objects (like, for example, bits of light called photons), it doesn't quite work.

It turns out that our current best theories about tiny particles are incompatible with relativity. Somehow, somewhere, we're still a bit wrong.

## Solving the problem

Partly, these mistakes are explained away by progress. As we get better at physics, at coming up with clever experiments and measuring the results more and more accurately, our theories slowly get better.

A good theory should always make predictions about things we can see happening (for example, if we double the weight we hang on it, our spring will double in length). If we see a prediction come true, that supports the theory. If we don't, it is disproved, and we have to try again. Every time we see what we expect, our confidence in the theory is increased, but it only takes one counterexample to destroy the theory. At least it does in theory.

Actually, when a physicist—let's call him X—gets a "wrong" result, science has a problem. One way to solve it is to adapt and improve the theory, without changing its basic rules, so that it predicts X's results, too. That way we don't have to throw away whole, useful theories just because that troublemaker X was up to his experiments again.

When it comes to springiness, we say that Hooke's law only holds for certain kinds of material, under certain conditions—to be precise, it holds when springs are springy. Pull your spring too hard and it reaches a point where it turns into a twisted piece of wire. Keep pulling and it will break. But breaking a spring doesn't break Hooke's law, which long ago stopped applying to that spring.

Simply improving a theory, setting limits and exceptions, is pretty hard work. One easy alternative is to accuse X of getting his results wrong and wonder why he gets so much money for his wrongheaded research. There's always plenty to argue about in science—for researchers that's half the fun of it.

## Revolting physicists

Sometimes, the little adjustments and strange results begin to mount up, as they did for Newtonian physics in the nineteenth century. Our best theories become a hodgepodge of exceptions and special cases, and arguments rage about unexplained results and experimental error.

At this point it takes a revolution to sort things out, to explain a host of problems, and to give us a whole new way of seeing the universe. The arrival of quantum physics was just such a revolution. Newton's universe was solid and precise and ran like clockwork. Quantum physics says that this is just how it looks to us, but that underneath it is flimsy and never completely predictable.

In the same way, some think that the current struggle to produce a unified theory may be a signal that relativity and quantum physics are missing something important, too. No one knows what the next revolution might look like.

## Fundamental physics

The hope of physicists is that these revolutions give us a deeper understanding of our world and how it works. Hooke's law is fine, but we now also have complicated theories (about how atoms move in a piece of metal and the law of conservation of energy) that explain Hooke's law, and what happens to a piece of metal when Hooke's law stops applying, how it stretches permanently and eventually breaks.

So there are laws that explain other laws. And somehow the explaining laws (which should be equally testable in experiments) are more important and deeper than the others. At the bottom of all the layers of explanation, we think that there may be just a few fundamental laws that will prop up all of physics and the universe we know.

Even now, all our best attempts at the deepest laws—which seem to explain almost everything that happens in the universe and to explain all the other laws we know—could be written, in mathematical symbols, on a single sheet of paper.

One goal of physics is to reduce that sheet to a single mathematical sentence that explains everything. Another is to get it exactly right.

## Putting it all together

The idea of fundamental laws leads to deep questions about what exactly physics does. Physicists tend to believe that there are real laws to discover, but they find it difficult to answer questions about what these are and how they work.

We'll come back to these debates, and what they may tell us about our place in the universe, at the end of this book, when we've had more of a look at the laws themselves.

For now, we'll note that physics tries to fit together a picture of the whole universe, which means that to understand one part you need to understand another, but to understand that part it helps to grasp the first. So don't give up if you don't get one part. You might understand after reading something else.

## Mad (and angry) scientists

It's also worth remembering that physics—like football, fashion, fishing, and even things we do that don't start with "f"—is a human activity. Physicists are people. They argue about their ideas, they make mistakes, and they do what needs to be done to make sure they have money to continue their work next year. Many of them work obsessively on their ideas and get very excited about them. And it's only with a dose of hindsight, when we forget the dead ends and the debates, that the story of physics seems neat and tidy.

Maybe another way to put it is that many physicists are, quite frequently, a bit nutty. Einstein, for example, liked to go sailing when there wasn't any wind ("for the challenge," he said). Newton, though, wasn't so much crazy as just very cranky (possibly because the mercury he worked with got to his brain). He went to extraordinary lengths to get his scientist friends to gang up and belittle some of his rivals.

Nikola Tesla, the gifted Serbian physicist and engineer who made huge advances in the study of electricity, probably had obsessive compulsive disorder. He insisted that the number of his hotel room be divisible by three; he didn't like to touch round things; and he kept saying his physics theories were right after he moved to the United States and ended up fixing roads because no one believed him. The thing is, he was (mostly) right.

## The nuttiest of them all

But the best mad physicist was an ancient Greek named Archimedes. It seems unlikely that he actually ran naked down the street shouting "Eureka!" (which means "I have found it!") because he was so excited after noticing that the water level rose when he stepped into his bath. But then again, in doing that he had just discovered the answer to an important question asked by his king, Hiero of Syracuse, and he was then able to prove that a crown the king had been given was a fake.

Archimedes is important because he reminds us that, in physics, we can all enjoy "Eureka!" moments.

We can all appreciate the everyday—but amazing—physics of rainbows, and make a good argument for our belief that the world is round even though it seems flat. And with a little effort (a bit of peace and quiet, accompanied by a stiff drink), we can even think through a thought experiment that demonstrates something as odd as Einstein's theory of relativity.

Read on, and see for yourself. Just remember, at all times, to keep your clothes on and your wits about you.

# 2

# Forces

*In this chapter, we'll find out how pushing and pulling change the way things move according to Sir Isaac Newton's three laws of motion; and how sometimes, no matter how hard you try, you won't get anywhere.*

OK. So let's get moving.

Four hundred years ago, as the scientific revolution started putting the whole world in a spin, Galileo, Newton, and other early physicists spent much of their time looking at how and why things move. And if forces were a good enough place for Galileo and Newton to start, they should be good enough for us, too.

They're also a good place to start because we intuitively have a feeling for forces and motion. They are part of the physics that we use and control every day: for walking, lifting, carrying, driving, and playing with hammers. And occasionally, forces and motion remind us how big and wild and implacable the laws of the universe are. For example: Don't play with hammers—you may drop one on your toe.

So we know that the faster we drive, the longer it takes to stop; or that when it's just you in the car, it goes like a rocket, but fill it with luggage and your extended family, and it would practically lose a race with a snail.

Sometimes, though, our intuitions about forces and motion mislead us. Let's look at an example.

## Float like a feather, fall like a hammer

We all know that something as light as a feather (like, for example, a feather) will fall more slowly than something heavy (like, for example, a hammer). Except, sometimes it won't.

In July 1971, when Dave Scott, commander of NASA's Apollo 15 mission, stood in the desolate moonscape and dropped a hammer and a feather at the same time, they fell together, and hit fine powdery moon dust together. And since it cost the American taxpayer around $150 billion (at today's prices) to get him there, we'd better hope we can learn something from that.

Of course, if you drop the feather on Earth, it doesn't fall straight down, but floats in the breeze. Actually, the breeze is the clue to what's happening. On the moon, there's no air to slow the feather. Taking the air out of the equation by doing the experiment in a near-vacuum on the moon is a great example of the way physics simplifies problems so that we can find out exactly what's happening.

Anyway, our experience with light things like feathers still makes it difficult to get a feel for the physics.

Somehow, even after all of Scott's effort, it still seems as if a 4-lb. (2-kg) weight will fall faster than a 2-lb. (1-kg) weight. Well it won't. And here's how to prove it in your head, using a thought experiment.

For this thought experiment, you'll require the following: three imaginary apples.

Imagine dropping three apples, all at once, from shoulder height to the floor. (It's easy: Just imagine you have three strong arms to hold them. And imagine you're really tall so that they take a long time to fall. See, you can do this physics stuff.)

So, all three apples hit the ground at the same time. Of course they did. Now what if we loosely tie one apple to another with a handy piece of imaginary string, and drop all three again. Again, all three crash down together. Of course they do. How could loosely tying two apples together make them fall faster?

Imagine making that piece of string shorter. Does that change the speed at which the tied apples fall? Of course not. Even shorter? Still no change. In fact, even if we get a whole roll of sticky tape and turn two apples into one big messy apple bundle, it won't suddenly start falling faster than its pristine comrade, even though that messy bundle is now twice as heavy.

So that's it, all proven—heavier things don't fall faster. Enjoy your apples.

And let's head back to the moon.


## The gravity of the situation

Have you ever seen videos of Neil Armstrong or any of the other lunar astronauts out on a moonwalk? Have you noticed their strange gait, how they lope around and execute huge two-footed kangaroo hops?

On Earth, heavy things and light things fall at the same speed (if the breeze doesn't get in the way), and the same is true on the moon, but there is an important difference between what happens here and up there. It turns out that the actual rate at which things pick up speed when falling (their *acceleration*, to use the proper physics word for getting faster) is less on the moon than it is on Earth. So even in a vacuum, it isn't how heavy things are that decides how quickly they fall, it's where you are when you do the experiment.

In fact, moon-walking astronauts could jump six times as high as they could on Earth. The reason why has nothing to do with Michael Jackson's famous dance move and everything to do with a force called gravity, which is what sticks us all to the surface of the Earth and makes everything right side up.

But what's a force?

## Feel the force

We can't see forces. But we can feel them and see their effects. The force of gravity keeps the Earth traveling around the sun and the moon traveling around the Earth, while other forces inside atoms hold them together and help create the matter that makes up everything around us. Forces make your car go faster and slow it down, push it around corners, and stop it from sinking into the road.

Pushing, pulling, turning, squeezing, and stretching all involve forces; we see objects move differently or change shape because of these forces every day of our lives.

> ## Newton's first law
>
> Isaac Newton summed this up in his first law of motion:
>
> *Things stay at rest, or continue moving in a straight line at a steady speed, unless acted upon by a resultant force.*

This is a beautifully simple law. But that doesn't mean you don't have to be careful with it.

## Direction makes a difference

OK. We're in that car at the start of the chapter. Lugging the family away somewhere nice on vacation. We're doing a steady speed of 67 m.p.h. (30 m/s), on a straight section of the highway, in the outside lane, probably with a long line of people in a hurry behind, all cursing and complaining and wishing we'd move over so they can break the speed limit and get wherever they're going 10 minutes earlier. But, hey, this is an important experiment, so they can wait.

We're traveling at a steady speed, in a straight line. Are there forces acting on our car?

Try rolling down the window and sticking your hand out. Phew! That's windy. So there's a pretty strong force (wind resistance) slowing the car down. But doesn't Newton's law talk about forces making things speed up or slow down?

True. It does. But there is one vital word we haven't examined yet.

*Things stay at rest, or continue moving in a straight line at a steady speed, unless acted upon by a **resultant** force.*

Wind resistance (along with a few other forces, such as friction acting on the wheels) is slowing your car. But the engine is working to overcome that resistance and keep it rolling at a steady 70 m.p.h. (113 km/h). And the result of these forces, one pushing forward, and an equal bunch pushing back, is no overall resultant force acting on your car. They cancel each other out and you keep up a steady speed.

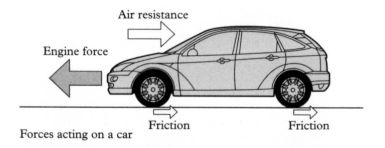

Air resistance

Engine force

Friction                    Friction

Forces acting on a car

Not only does this tell us how clever Newton was to put the word "resultant" in, it also points to an important aspect of forces: They have direction as well as size, or as physicists like to say, forces are a *vector* quantity.

*Scalar* quantities, like volume, on the other hand, are one-dimensional. A quart of milk is purely a quart of milk. Add it to another quart of milk in a big jug and you have two quarts of milk; there's nothing else to think about when add those up.

*Vectors* are more complicated. If I'm pushing a car with a force of 1 Newton (that's right, Sir Isaac was such a genius, they named the unit of force after him), and you're pushing with a force of 1 Newton, neither of us is pushing very hard. And, more importantly, because I haven't told you which way we're pushing, it's not clear whether there is a force of 2 Newtons acting on the car, or none at all.

If we're both pushing in the same direction, you can add our pushing power: there is a 2-Newton force on the car, and if we're lucky, we might move it. But we could be pushing in opposite directions, and the resultant force would be zero. And in between those two extremes we could be pushing more or less in the same direction, or almost in opposite directions, and then we could figure out the exact angles of our pushing and ask a passing physicist to calculate the exact size and angle of the resultant force on the car.

## Balancing acts

Remember our floating, blown-around feather? Well, we should now be able to understand why, even though on the moon it falls as quickly as a hammer, on Earth it doesn't.

Feathers are, light, flat, and wide, so wind and air resistance have a large area to push on. This means that, as a feather falls on Earth, air resistance quickly creates a force that is big enough to balance the small force due to gravity that's pulling the feather down. And once those forces are balanced, the feather will float down at a constant, slow speed.

Hammers, on the other hand, have a small surface area relative to their heavy weight. So they can fall much faster before the forces of gravity and air resistance balance out.

---

### Terminal velocity

Once gravity and air resistance have balanced out, there is no resultant force on an object. No resultant force means no acceleration (remember Newton's first law!), and so the object will continue falling at a constant speed. This speed is known as *terminal velocity*; in other words, the final velocity it reaches.

---

On the moon, of course, there's no air resistance. The feather and hammer fall at the same speed because the force of gravity is all that's acting on them.

You can clearly see the effect of air resistance in action when you watch a video of groups of skydivers chasing each other and joining hands as they plummet toward the ground. They are experts at changing their size and shape, and therefore the air resistance acting on them, so they fall more slowly or quickly. In fact, they can vary their speed from about 125 to 200 m.p.h. (55 to 90 m/s).

## When forces balance

Floating and flying are also good examples of balancing acts. An aircraft stays up because the lift that its wings produce is enough to balance the pull of gravity. Similarly, boats float when the upward push of the water—which is due to an effect called buoyancy—equals the pull of gravity.

# Archimedes' principle

Archimedes discovered that:

*Any object, wholly or partly immersed in a fluid, is buoyed up by a force equal to the weight of the fluid displaced by the object.*

So if you gently float a small boat in a tub of water filled to the brim, and catch all the water that flows over the edge when it is displaced by the boat, you'll find that the weight of the boat is the same as the weight of the water you caught. Neat, huh?

Buoyancy is the reason we feel so wonderfully light and maneuverable when we swim. Even a diver with a heavy tank and a weight belt is so buoyed up by the water that she weighs almost nothing as she cruises around the coral reef.

Buoyancy also lifts hot-air and helium balloons in the air. Heated air and helium gas have lower densities than normal cold air. This means they are lighter than air, so a balloon filled with hot air or helium displaces a weight of air greater than its own weight. According to Archimedes' principle, this creates an upward force on the balloon.

This makes it clear why hot-air balloons have to be so big. A rough calculation shows us that, because air weighs only 1.2 kg per cubic meter, you need to displace at least 80 cubic meters of it to make sure that the displaced weight is greater than the weight of your average hot-air

*(continued on page 34)*

## What Archimedes told us about density

Buoyancy is also useful for finding fake crowns, as Archimedes is said to have demonstrated soon after his Eureka-shouting, running-around-naked antics more than 2,000 years ago. All you need is a little understanding of density, to go with what you know about buoyancy.

The density of an object is its mass divided by its volume. It's a measure of how much something weighs per 1 cubic meter lump (or sometimes in pounds per square inch).

| Material | Density in kg/m³ |
|----------|------------------|
| Air around us | 1.2 |
| Water | 1,000 |
| Lead | 11,340 |
| Gold | 19,300 |
| Core of the Sun | 150,000 |
| Neutron star | $1 \times 10^{18}$* |

*That's 1 with 18 zeros after it. Be careful where you drop that lump of neutron star. It's super heavy. Even a teaspoonful would weigh 10 billion tons on Earth if you could get it here, which you can't.

Archimedes' task was to find out whether a crown, given to his king, Hiero II of Syracuse, was really made from gold. First, he borrowed a chunk of gold from the king that weighed the same as the crown. Then he hung the gold from one side of a set of scales and the crown from the other, so they balanced. And finally, while it was still hanging there, he dipped the crown and golden chunk into a bath of water.

If the crown, like the chunk, had been gold, the scales would still have balanced, because they would both have displaced the same amount of water (and therefore felt the same upthrust). But, because the crown was a fake, it was less dense. That meant it had to be bigger to weigh the same amount as the gold. And because it was bigger, it displaced more water and was buoyed up more than the gold. So the crown rose and the gold sank (along with, it would seem, the fortunes of the stingy crown-giver).

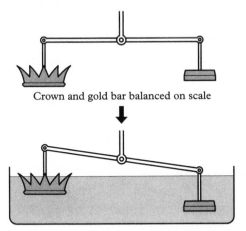

Crown and gold bar balanced on scale

Once in the bath, the gold bar sinks as it is more dense

Checking that a coin is real gold the Archimedes way is smart because, as you can see from the table above, it's more dense even than lead, so it's difficult to fake.

balloonist (who weighs around 80 kg). And to take up 80 cubic meters of space, a cube of air would need to measure 15 ft. (4.5 m) along each side.

Since we displace air, too, buoyancy is also acting on us. In air, we weigh about 0.1 percent less than we would in a vacuum. That is, the atmosphere gives us a weight reduction of almost 3 ounces (about 80 grams) for the average-sized adult! And we don't have to diet to get it.

## Speed and velocity

Let's get back into our car, still hurtling down the highway. But now, while we're roaring along at 65 m.p.h. (30 m/s), a side wind hits our car.

This new force isn't balanced by the friction pushing the other way (because you are caught up in the physics of the moment and you've let go of the wheel). So now there is a resultant force on the car, and it changes direction. It accelerates across the lanes of the highway, and a dozen cars have to swerve out the way, their drivers shaking their fists.

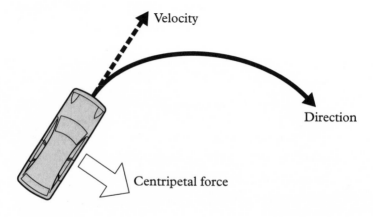
Velocity

Direction

Centripetal force

To take account of this, in physics, we talk about velocity rather than speed. Velocity is a vector quantity: Like a force, it has a magnitude* (the speed of the car) and a direction (at this point, veering dangerously across the highway). We tend to think of acceleration as getting faster, but in physics it means any change in velocity, so a change in direction counts as an acceleration.

For an object to move in a circle at a constant speed, you need a centripetal force: one that is continually changing direction so that it is always acting toward the center of that circle. Does that sound complicated? It isn't, actually. Swing a lasso around your head, and the rope is just right to provide a force toward the center of the circle (your hand) and to keep the hoop moving round and around (until you let go and the hoop, with no resultant force acting, flies off to rope in the nearest pedestrian). In a similar way, gravity, which always acts toward the center of an object, keeps the planets going around the sun at a constant speed.

Car seats are pretty good at this, too. The force you and your passengers feel as the car swings across the highway pushes so that instead of continuing straight down the highway at 65 m.p.h. (30 m/s), you all stay in the car and turn with it...

Look out! You're gonna hit that truck!

## Thinking time

We're all so used to zipping along highways that we hardly think about how quickly we're really traveling.

*Magnitude is just a fancy physics word for "size."

And highways are designed for fast driving: There are few features to distract us, bends and hills are long and gentle so we can continue quickly and see a long way ahead, and signs give us lots of warning so we're prepared for what's coming up.

Try doing 65 m.p.h. (about 30 m/s) in town and you'll soon get a feel for how fast the highway speed limit is. Or rather, don't. Have a quick look at the table below instead and stay safe.

| | |
|---|---|
| Speed of light | 300,000,000 m/s |
| Speed of sound | 300 m/s |
| Car on highway | 30 m/s |
| Person jogging | 3 m/s |
| Child crawling | 0.3 m/s |
| Snail | 0.03 m/s |

Thinking about your car's speed in unfamiliar units helps you come to grips with it. Every second, your car is traveling 98 ft. (30 m). Remember, that's 30 long paces. Every second.

When you look at it like this, it's hardly surprising that even a driver who is concentrating can travel almost 150 ft. (45 m) before slamming on the brakes after they notice that slow truck bumbling along in the inside lane. That's because it can take the human brain 1.5 seconds to realize, hey, that's a slow truck, I'm gonna hit it, to then decide to brake, and finally, actually hit the brake pedal. And that's an alert driver! If you're thinking deeply about physics, you might take even longer.

The upshot? Nothing. We're all in a hurry. But if we were sane we'd be scared witless every time we drive. Especially when you consider that so far we've only thought about distance...

## Screeching to a halt

Brakes on? Good. But you aren't slowing down very quickly, are you? There was a reason I told you the car's packed with luggage and your extended family. Didn't you notice when we pulled away from those lights back in town that those snails outpaced us for a bit?

As we started to see earlier, stronger forces cause greater accelerations, for any given object. The relationship is proportional. Double the force, double the acceleration (or, when you apply the brakes, the deceleration). But what if you're using different objects? Well, then you'll find that heavier things accelerate more slowly. Double the mass (mass is a bit like weight—we'll come to it in a moment) and you *halve* the acceleration.

Plugging some numbers into the equation should make it easy to understand the relationship between these three quantities. If our brakes are trying to stop the car with a force of 7,200 Newtons, and our car has a mass of

### Newton's second law

This is Newton's second law of motion. Physicists usually sum it up by writing:

*Force equals mass times acceleration*
*or*
$F = ma$.

(*ma* has nothing to do with your mom but is instead a short way of writing *m times a*.)

1,200 kilos (typical for an empty car), our deceleration would be 6 m/s every second, because

*7,200 N = 1,200 kg x 6 m/s².*

But if we add in four adults and their luggage (*1,200 kg + 400 kg = 1,600 kg*), we get: *7,200 N = 1,600 kg x 4.5 m/s².* Our deceleration is now only 4.5 m/s.

Physicists like to work with numbers and mathematical equations like this, but learning equations isn't the same as learning physics. If *F=ma* is all Greek to you, don't worry. You can still understand forces and how they make things move. You just don't know your a from your Ω.

## Everybody's equal (and opposite)

So, where were we? Safely stopped on the shoulder of the road? Good, good. And luckily, there's Newton's third law to keep us that way.

Remember gravity? It's pulling the car toward the center of the Earth. And the equal and opposite force? Well, it's pulling the Earth toward the center of the car! Of course,

---

### Newton's third law

This says:

*If a force acts upon a body, then an equal and opposite force (of the same kind) must act upon another body.*

## When golf balls go awry . . .

Newton's laws of motion apply perfectly to the world of golf. We're all aware that a golf ball moves when it is hit by force. However, there are outside forces that keep a golf ball from moving in its original direction forever. A ball may have a straight path when the club hits it, but then gravity pulls the ball toward Earth and can keep it from going straight. Air resistance—a form of friction—then slows the ball's velocity as it speeds through the air. Once a golf ball connects with the ground again, it slows down even more because a grassy or sandy surface creates more friction with the ball than air.

The reaction of a golf ball when a club hits it is usually predictable. When a force is applied to the back of a ball with a club by swinging—an action—the ball zooms down the fairway—a reaction. But golfers become extremely frustrated when the desired action and reaction doesn't happen. And when a golfer does not hit the ball squarely with his club, things can go awry—a slice, a hook...yes, sometimes, even a broken window.

the mass of the Earth is $5 \times 10^{21}$ times greater than that of your car, so the effect on the Earth isn't noticeable.

Here's another example of the third law. Remember when we were braking hard as we tried to miss that slow truck? Friction applied a force of 7,200 N to our car that slowed it. And the equal and opposite force? Well, it's 7,200 N acting in the opposite direction on that bit of the surface of the Earth known as the highway. So, although the effect isn't noticeable because the Earth is so massive, our braking does actually affect the way it rotates.

Anyway, enough of these death-defying near-misses. We need to save our energy for something much more important, called—guess what?—energy.

# 3

# Energy and Power

*In this chapter we'll talk about all kinds of energy: energy that things have because they're moving, because they're where they are, or even because of what they're like inside. We'll also talk about power, or putting that energy to use. And finally, we'll put it all together and find out how the physics of energy can save us from going out with a bang.*

Energy doesn't like to sit around, which is hardly surprising, given that it's so energetic. Energy is always here, there, and everywhere; always doing this and that; always changing into something new. It's like the trendiest party animal in the universe. Besides propelling us down highways and into physics experiments, energy keeps us warm, grows our food, and helps us see. It even gets us up in the morning and makes the toast.

All our energy comes, initially, from the sun. The sun is one great big long-lasting nuclear explosion, and it's chucking solar energy at us at a fantastic rate. A third of what comes our way is bounced straight back off the Earth's atmosphere and the clouds and goes back into space, yet our planet still absorbs more energy from the sun in *1 hour* than humans burn in *10 months*.

So the sun warms us and the planet (the energy that it sends becomes thermal energy). It makes the plants grow, turning solar energy into chemical energy through photosynthesis. We eat some plants, turning their chemical energy into all kinds of other energy, such as kinetic energy (energy you have because you're moving), as we bounce out of bed, and sound energy as we shout at whoever's in the bathroom to get a move on. The sun's energy also evaporates water from the sea and turns it into clouds, and generates the air currents that turn into wind.

Some plants nourished by the sun millions of years ago have been turned into coal and oil. We dig and burn them; they warm us and light our way when the sun's hiding behind clouds or on the other side of the planet. Or we burn them and they shoot us off on vacation, fueling our airliner as we head someplace where the sun's still shining.

## Counting the calories

If you're on a diet, you count calories*, but if you're a physicist, you measure energy in Joules.

*1 calorie = 4.2 Joules.*

*The calories you count are kilogram calories (or kcals), but in most other contexts, we are actually talking about gram calories. A calorie is the amount of energy needed to heat some water by 1°C, and the "gram" or "kilogram" refers to exactly how much water, so a kilogram calorie is 1,000 gram calories. (There is also the rarely used pound calorie.)

Joules are named after James Joule, a wealthy brewer from Manchester, England, who was fascinated by electricity. As a child, he experimented by giving his brother and some of his servants electric shocks. But in the 1840s, when he'd grown up, he made himself useful by looking at how he could use the newly invented electric motor instead of steam engines in his brewery. From there, he discovered lots of interesting things about how electrical and mechanical energy is converted into heat.

Eventually it was decided that 1 Joule would be the amount of energy expended by a force of 1 Newton if it moved an object 1 meter. (They try to keep it simple, don't they?)

So in real life, a Joule is the energy it takes to lift a small apple 1 meter straight up. Or the kinetic energy of a tennis ball lobbed gently over the net. But there are around 720,000 Joules of food energy (carbohydrates and alcohol) in a pint (500 ml) of beer, which is probably why you're going to have to try something more strenuous than the apple-lifting workout if you want to get rid of that beer belly.

## What a waste

However, converting energy is rarely straightforward. Modern experiments on rowers show that for every 5 Joules of energy they burn, only around 1 Joule is actually used to move their oars. The rest goes in heat, in sweating and swearing and making pained expressions for the camera.

# Energy conservation

Joule's work on energy conversion led to the discovery of the law of conservation of energy, one of the most basic and important in all physics. Once you know that energy is converted from one kind to another, careful measurement can show that all of it is always converted, that energy is never created or destroyed. Then you have a very important law: the first law of thermodynamics.

It says:

*The total amount of energy in a closed system remains constant.*

So our universe, the biggest closed system we know, is just as energetic now as it was when it first exploded into existence. And when Mr. Joule gives an apple 1 Joule of potential (height) energy by lifting it up, all that energy becomes kinetic energy as it falls, and then heat and sound (splat!) energy when it hits the floor.

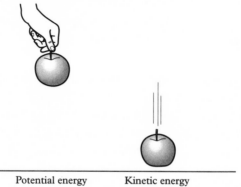

Potential energy        Kinetic energy

And it's not just Olympic gold medal winners who waste energy. Something similar is true of all us when we burn food energy. As a Victorian gentleman, Mr. Joule can surely lift an apple without swearing, but he still wastes 4 Joules of energy doing it.

Most energy conversions are, like Mr. Joule's, very inefficient. Even modern low-energy lightbulbs only turn half the electrical energy they consume into light. The rest is, again, wasted as heat.

In old-fashioned, conventional lightbulbs, only 10 percent of the energy is turned into light, with the rest warming the room. But even that looks efficient compared to some other energy conversions.

It has been calculated, for example, that for every Joule of energy available in Mr. Joule's pint of beer, at least 10,000 Joules of energy have been used in making that beer. That energy is applied mainly in the form of sunlight

## The definition of work

When Mr. Joule picked up that apple, he did some work. Not much work, agreed, but certainly a physicist would call what he did work. That's mainly because, in physics, work is defined as "energy transferred during a process".

When that apple is lifted, it gains 1 Joule of energy. We know that because it has a weight of 1 Newton (so it took 1 Newton of force to lift it) and because we know it moved 1 meter, and *work done (or energy gained) is force times distance moved in the direction of that force.*

If we drop the apple, that energy is transferred again; gravity does the work to speed it on its way to the ground.

growing the hops and barley, and in the heating required at various stages as the beer is brewed.

## Watts and work

It isn't very exciting that Mr. Joule can lift an apple to the not-very-high height of 1 meter. But what if he can lift that apple quickly? What if every second, he lifts an apple a meter, so he's lifting 60 apples a minute, and 3,600 an hour? Well, then he'd be doing 1 Joule of work per second, which means his power output is 1 watt, for no other reason than that is what a watt is defined to be.

The watt (W), which is the unit we use to measure power, or how quickly something does work, is named after James Watt, a Scottish inventor whose redesign of the steam engine in the 1700s gave it the power to be the workhorse of the industrial revolution.

## You have more energy upstairs

As you will have realized from all that apple lifting, the higher something is, the more gravitational potential energy it has. That applies to you, too. The higher you go, the greater your potential energy is. In fact, your potential energy is directly proportional to your height off the ground. If you're eight stories high in a hot-air balloon, you have 8 times more energy than when you were only upstairs at home.

The amount of potential energy is also dependent on your weight. If you're heavier, you don't fall faster, but you will make a bigger splat when you hit the ground.

You can do the thought experiment. We already know that if you irresponsibly drop from your balloon two apples joined by a loose piece of string, they don't fall faster than one apple. But, compared to one apple, those two do make double the mess and double the bang when their potential energy, after becoming movement (kinetic) energy as they fall, turns into splat energy as they hit the pavement and explode, lightly coating surprised passers-by with apple juice. The law of conservation of energy says that splat energy had to come from somewhere, so it must have come from their potential energy, which became kinetic energy.

So it's clear that an object's gravitational potential energy is directly proportional to its weight (eight apples have 8 times as much potential energy as one, which will be released as 8 times as much splat energy when they hit the pavement).

It is weight that is important here, not mass. On the moon, in one-sixth Earth's gravity, those apples would fall more slowly and splat much less. How much less, we don't know, unfortunately. Thought experiments only get you so far, and no Apollo astronauts threw apples out the window of the lunar lander.

The law of conservation of energy applies in this experiment because our falling apples are very close to being a closed system. Only a tiny amount of energy is lost to air resistance. At a point when the falling apples are just a fraction of an inch (a few millimeters) above the pavement, and just before they splat, they have lost all their potential energy: It has all been turned into kinetic energy. A millisecond later and all that kinetic energy has become splat energy.

Frankly, though, finding the kinetic energy of falling apples is not that useful. What about the kinetic energy of a speeding car?

## Car crash equations

When you're driving down the highway and brake hard to avoid a slow truck or bus, your brakes do work (in the physics sense) to slow you down.

Of course, if your brain was working, you'd have remembered all those stopping distances you learned in order to pass your driving test, you would have left a bigger gap between you and the car in front, and you wouldn't have had to brake so hard in the first place. But then again, I'm sure you forgot those facts about stopping distances as soon as you had your shiny new licence and 15 friends in the back of your parents' station wagon for

| | | |
|---|---|---|
| 20 m.p.h. | Total = 12 m | |
| 6 m  6 m | | |
| | | Thinking distance |
| | | Braking distance |
| 30 m.p.h. | Total = 23 m | |
| 9 m    14 m | | |
| 40 m.p.h. | Total = 36 m | |
| 12 m      24 m | | |
| 50 m.p.h. | Total = 53 m | |
| 15 m         38 m | | |
| 60 m.p.h. | Total = 73 m | |
| 18 m            55 m | | |
| 70 m.p.h. | Total = 96 m | |
| 21 m             75 m | | |
| 80 m.p.h. | Total = 120 m | |
| 24 m                96 m | | |
| 90 m.p.h. | Total = 153 m | |
| 27 m                126 m | | |

a celebratory drive around town. Of course, you are now mature and fascinated by physics, so stopping distances are worth a closer look.

If you double your speed, from say 20 m.p.h. to 40 m.p.h., your stopping distance is quadrupled (from 20 ft. up to 80 ft., or from 6 m to 24 m). Double your speed and you quadruple your stopping distance. Triple it (from 30 m.p.h. to 90 m.p.h.) and your stopping distance increases *ninefold*. This is obviously one factor that makes speed so dangerous. But why does it happen?

Well, kinetic energy has the answer, but only if we know a bit more about friction, the force that makes your brakes work. (And it's just as well they do work. Imagine if you ran head on into a slow truck, or worse, a brick wall. Well the good news is, you don't have to imagine it: we'll get to that, too.)

## Get a grip

Friction is a force caused by the resistance of surfaces as they slide over each other. Without it, we'd all be slipping around like non-skaters on an ice rink.

Scientists used to think friction was caused by the bumps and lumps of the surfaces as they knock together. But it seems friction is actually mostly caused by chemical bonding between the moving surfaces: It's more like sticky tape being dragged along a surface, rather than sandpaper.

Anyway, friction causes motion energy to be transferred into heat and screeching sound energy. Or, as a physicist might say, the heat and sound cause the frictional force (because the transfer of energy away from motion is what causes the force to be felt). Confused? Don't worry—so are scientists. There's still a lot to learn about friction. But we can all agree that friction is greater the harder you push the rubbing surfaces together, so heavier things feel greater friction: It's easier to slide a single armchair around the living room than a great big sofa full of couch potatoes.

Friction also depends on the kinds of surface involved. It's not just fashion that encourages us to wear rubber-soled tennis shoes rather than leather-soled loafers on the tennis court: we wear tennis shoes because they grip better.

Anything that comes between surfaces can affect friction. For example, fluids, such as oil and water, reduce it by keeping the surfaces apart. This is useful inside your car's engine (it stops energy from being wasted as your crankshaft turns), but it's dangerous outside when the road gets wet or oily.

The work the brakes do, according to the law of conservation of energy, converts all the kinetic energy of the car into heat, through the friction of the brake pads on the brake discs.

For a car braking hard from highway speeds, that's a lot of energy to convert. During hard braking, each brake heats up with the power of a small electric fire, and if you hit the pedal hard a few times in a row, that heat can build and build until your brakes stop working properly (which is why signs are sometimes placed at the top of steep hills encourage you to select a low gear and use your engine to slow you).

Now, here's the clever part. We know that all our car's kinetic energy has been converted into heat as the brakes did work to slow it. We also know, from the "work" equation we looked at earlier, how much energy that is: It is

## Slip-sliding away

There are two kinds of friction: static and dynamic. Static friction is the force you feel when you're walking carefully on an icy pavement. There's quite a lot of grip there: So long as you move gently, your feet don't slip. But as soon as they start to slide, and the much lower force of dynamic friction takes over, you're on your backside. This is one reason why cars have anti-lock (or ABS) brakes. They are designed to release before your wheels slide on the road, and lower, dynamic friction takes over.

given by the braking force multiplied by the distance moved. And we know something odd about that braking distance, namely that it quadruples if our car goes twice as fast. Therefore, the kinetic energy of our car must also quadruple if we go twice as fast. So, kinetic energy must be proportional to the speed of our car squared.

See how $v$, the speed (or velocity) of the object, is squared, so that if we double the speed we get 4 times the energy ? And if you're really good at math, you can actually derive this formula from the work equation ($F$ *times* $d$). That's interesting, isn't it? There's something about our math, and the way the universe works, so that when we do clever mathematical things to our work equation (*force times distance*), we get our kinetic energy equation. That's the start of how mathematics and physics go hand in hand.

Even better: Now that we have the equation, you don't need to look at government figures to figure out your stopping times. All you need to do is one experiment, in a deserted parking lot. Measure how quickly you can stop from 20 m.p.h. (32 km/h). Then you can quickly fill in your own table. If you doubled your speed, to 40 m.p.h. (64 km/h), your stopping distance would quadruple. If you multiplied it by 3.5, to get 70 m.p.h. (113 km/h), your stopping distance should increase by $3.5^2 = 12.25$ times.

## Watch that child

Government figures also tell us that, four times out of five, a child hit by a car traveling at 40 m.p.h. (64 km/h) will die. But if the car that hits a child is traveling at 30 m.p.h. (48 km/h), four times out of five the child will survive.

Drivers who don't like speed cameras or speed bumps take issue with these figures. How do we know, they ask, the exact speed of a car when it hits a child who has run out into the street? It certainly seems unlikely that anybody has been throwing children in front of cars traveling at various speeds to see how many survive.

Well, even if you don't believe government figures, and skid mark data, and experiments that were done on dead bodies and animals, and the stories told by the kinds of injuries children receive when tossed into the air by a carelessly driven car, physics might convince you to slow down. Firstly, there are those thinking and stopping distance figures we've already looked at. At 30 m.p.h., if a child stepped out 82 ft. (25 m) ahead of you, you'd hit him or her at 19 m.p.h. (31 km/h). At 40 m.p.h., you'd hit them twice as fast—at 38 m.p.h. (61 km/h).

You can also look at the forces on the child's body. Let's say a child is struck by a car, and carried down the road

# Crash test hero

Crashing into children is horrible; but crashing into other cars is—apart from being dangerous, noisy, and expensive—quite exciting. That's why we like bumper car rides at amusement parks—and why some people have an unhealthy interest in accidents.

Starting in the 1940s, Dr. John Stapp, a U.S. Air Force officer, did lots of research into the forces that human bodies can survive, by appointing himself crash test dummy and driving a rocket-powered sled into what was, basically, a wall. In this way, he proved that the human body can survive accelerations 40 times stronger than gravity—so long as they act for just a short fraction (0.05) of a second.

Stapp played the dummy, but he was sensible enough to wear a seatbelt. And, although his crashes broke numerous bones and caused permanent damage to his eyesight (because his retinas became detached), he survived to discover a lot of useful information that helped others design aircraft ejector seats and car seat belts.

Later, his campaigning was a major factor in making seat belts a compulsory feature on U.S. cars. (He also came up with Stapp's law: "The universal aptitude for ineptitude makes any human accomplishment an incredible miracle." It explains why so many people didn't and still don't bother using those belts he worked so hard to put in their cars.)

by it. That means they have been accelerated very quickly to the same speed as the car (because hitting a 40-pound child with a 2,600-pound car won't slow the car down much). But we know from the kinetic energy equation that the force involved in accelerating something is proportional to the square of the speed. So, although our 40 m.p.h. car is traveling twice as fast when it hits the child, the force that acts on the child's body is 4 times greater. And human bodies are very soft compared to cars.

## Smashing fun

So we've looked at stopping times and seat belts. But what else can physics tells us about how to make cars safer for humans hell-bent into crashing them in thought experiments?

Let's say we build a really strong little car. You can smash it, head on, into a brick wall, and it'll hardly get a dent. However, if we think of the work equation and kinetic energy again, it's clear that, when you hit that wall, the fact that your car stops very quickly means that the force required to stop it—and you—is very large.

Let's do the math.

From the kinetic energy equation, a 175-lb. (80-kg) imaginary person named Bob, who has for some unknown reason volunteered to be in our little car, is driving it at 60 m.p.h. (27 m/s). As a result, he has:

$$\tfrac{1}{2}\ mv^2 = \tfrac{1}{2} \times 80 \times 27^2 = 29,000\ \textit{Joules of kinetic energy.}$$

Now, let's say, as Bob drives head-on into a wall, that his seatbelt stretches enough to let him move 4 inches forward (that's 10 cm, or 0.1 m) as it stops him from crushing his skull on that hard little steering wheel. Then the force required to take away all his kinetic energy is given by:

$$KE\ (29,000\ Joules,\ in\ this\ case) =$$
$$F \times d\ (force\ times\ distance) = F \times 0.1$$
$$So\ F\ is\ 290,000\ N.$$

For an 175-lb. man, this is an acceleration 360 times bigger than gravity.

Well, that's no good. Bob's little car might still work, and his body might still be strapped in, but after experiencing a force like that, his brain has been squished into soup as it was stopped by the inside of his skull. Stapp was pretty much OK after acceleration 40 times stronger than gravity; a racing driver named David Purley almost lived after he crashed and experienced 180 G. But at 360 G (in other words, 360 times the force of gravity), you've got no chance.

To survive, you need to take longer to stop. So what modern car engineering has done is to slow things down by adding, to the front of our car, a 3-ft.-long (1 m) crumple zone: an area that is strong enough to take some of the sting out of an accident by crumpling. This converts some of the energy of the collision into heat and sound, and making it take longer for the car to slow, so that the force on you is reduced.

Add this to our seat belt stretch, and you now have 10 times as long to stop as Bob did (up from 0.1 m to 1 m, because the crumple zone shrinks from 100 cm to 10 cm in the accident).

Now, when the car stops, the 10 times the distance to stop means the force on you is reduced to a tenth; which means the acceleration is a tenth, which means it is only 36 times the force of gravity. And Colonel Stapp proved you could survive that.

Hooray. The numbers work. So, in you go, but do drive carefully. Oh, and do buckle up. Steering wheels and dashboards are very efficient stoppers. They'll bring you to halt in about ½ inch (1 cm). That's 10 times quicker than a seatbelt: enough to make even a 30 m.p.h. crash fatal. And enough to make it unlikely you'd ever come to grips with the momentous stuff we've got coming up next.

# 4

# Momentum

*In this chapter we'll find out about momentum and inertia, and how physics shows us that, once something starts moving, that movement never stops. Then we'll go around in circles, and discover how spinning tops and gyroscopes rely on similar principles. To finish up, we'll talk a bit about torque, the force that makes big engines fun.*

In the martial arts movie *Crouching Tiger, Hidden Dragon,* the heroes can break the laws of physics, which makes for all kinds of spectacular floating and flying as they fight. At one point the rebellious young heroine, Jen Yu, is attacked as she's having lunch. A young tough guy runs at her, but, without standing or bracing herself, she blocks him with her left hand, and he bounces off her. All the while, she holds her cup of tea in her right hand and doesn't spill a drop.

Her move, though handy, is obviously impossible. But why? Her attacker has just as much kinetic energy after she blocks him as before. He rebounds, so energy is conserved, just as it should be. Yet some law of physics must have been broken—otherwise a smart flyweight boxer would by now have taken a few classes in Jen Yu self-defense and become heavyweight champion of the world.

It's clear, then, that the laws of energy alone cannot show why Jen's heroics are impossible. For a start, energy lacks a sense of direction—it is a scalar quantity, like mass and volume. It has size and nothing else, so it can't explain why it's odd that Jen wasn't knocked backwards (and into next week).

But lack of direction isn't energy's only failing when it tries to describe motion. To discover more, let's try a thought experiment on a glacier: Let's try to stop it.

## Hold the ice

A glacier is a massive, slow-moving river of ice, so it won't mind us giving it a shove back the way it came. Come on, push as hard as you can. Any luck? How about if you bring in a fleet of tow trucks? Well, it's still hopeless. Even hills and rocks can't stop a glacier: It plows right on through whatever is in its way, slowly carving out massive U-shaped valleys as it goes. The forces involved are tremendous.

But what if we string a giant waterwheel-kind-of-thing over a glacier, and try to convert its kinetic energy into electricity, as we could with a river. Can we solve the energy crisis using glaciers?

Well, let's continue with our thought experiment and find out. Let's say our glacier is a 1 cubic-kilometer (0.6

miles cubed) block of ice. Its mass is roughly $10^{12}$ kg (1 followed by 12 zeros) because water weighs 1 kg per cubic meter. And let's say it moves, like the Byrd Glacier in Antarctica (the biggest in the world), at a speed of around 0.00001 m/s. Then its kinetic energy is a not-so-whopping 50 Joules. Convert that all to electricity and you get a low-energy bulb lit for a few seconds.

Again, energy isn't giving us the full picture. Glaciers are unstoppable—but they have about as much energy as a student physicist after a night's overindulgence on Mr. Joule's brew (see chapter 3).

What we need is something that describes unstoppable glaciers and lunging bad guys. That something is called *momentum*.

## Don't stop now

Momentum is something we all intuitively understand: It's why you can't stop a car by giving it a shove; it's what makes pool and billiard games worth playing; and, as you'll have guessed by now, it comes up in a law of the universe that makes Jen Yu self-defense impossible.

The idea of there being something in a moving object that keeps it moving has been around since Roman times. But it wasn't until Newton came along that it was properly, mathematically useful. This is hardly surprising, as momentum is intimately tied up with Newton's laws of motion, especially the third law, as we'll see when we discover that flaw in Jen Yu self-defense.

## The momentum equation

Momentum is calculated by multiplying mass by velocity, and so is given by the formula:

$P = mv.$

So our glacier from earlier can have huge momentum even though it is moving slowly. (Its momentum, 10,000,000 kg m/s, is big: about the same as that of a giant car transporter carrying 300 cars quickly down the highway.)

Momentum is why oil tankers take 20 miles (32 km) to stop. It's a measure of the "quantity of motion," so it is proportional both to the mass of an object and to its speed.

## A load of balls

As physicists, we like to simplify things to find out what's really happening. So instead of trying to find a Jen Yu martial artist and a cooperative thug, we can recreate the scene with pool balls. But if we're going to do experiments on a pool table, we want to carry them out on a physicist's *ideal* pool table: one that's perfectly smooth, not bumpy or beer-stained. Then again, ideal pool tables are in short supply. So you may as well find a friendly bar or pool hall, get yourself a drink, and settle down for some physics.

First up—to play the part of a relaxed and lunching Jen Yu—choose a ball that's blue. Then, entering from stage left, playing the thug, a gently rolling red. (We won't use the white ball because, on most pool tables, it's a different size and weight than the others, and that makes everything less ideal.)

Now, red hits blue, full ball, right in the middle. What happens? Well, unless you muck things up by putting spin on the shot, the red stops and blue keeps going—and it keeps going at exactly the same speed, and in the same direction, left to right, that red was moving.

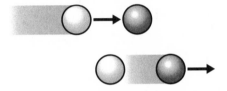

How momentum works on pool balls

Try it again. And again. Try it in a different pool hall. And another. The result will always be the same (so long as you get the collision smack dab in the middle of the blue ball).

Now that makes sense. Again, energy is conserved, but now our new quantity, momentum, is, too. All the energy and all the momentum on our pool table has been transferred from red to blue.

Newton's third law—the one about equal and opposite forces—has been obeyed, too. As the balls collided, the amount that red pushed blue equalled the amount that blue pushed red, and that amount was enough to stop red and send blue on its way.

The amazing thing is that what you have discovered about momentum, right there in front of you on the pool table, is one of the most basic laws of the universe.

## Exactly right

Energy and momentum are so important that they are mentioned in a bunch of laws that physicists are pretty much absolutely, completely, honest-we-haven't-got-these-wrong sure about. As far as we know, these laws have never been broken. They applied in the beginning, right after the big bang. And they will apply forever, until the end of time.

They are called the exact laws.

The exact laws are all conservation laws. They say that the total amounts of certain properties in the universe are always conserved, so that these properties can be neither created nor destroyed. (More usefully for those of us who aren't gods, they also apply to small closed systems like our pool table, if we're careful.)

We've already seen the law of conservation of energy. Now that you've done the experiment with the pool balls, you have discovered the momentum version for yourself. It says:

*The total momentum in a closed system remains constant. It cannot be created or destroyed.*

Energy and momentum are always conserved in our universe simply because of its underlying shape, which is the same whichever way you go, and because time ticks on constantly, without changing the laws of the universe.

(The other exact laws describe the conservation of slightly more complex properties, such as electric charge, and much more complex properties, such as color charge (nothing to do with the price of red paint), weak isospin (nothing to do with a gentle wool wash) and—unluckily for gamblers—probability.)

So conservation of momentum is there for racing drivers who want to slingshot out of corners, and for football players who need to flatten their opponents. It's there for pool players and for people who like to poke holes in martial arts movie fight scenes.

What's more, and like we saw in the last chapter with energy, momentum can be simply and beautifully useful in everyday physics, because it is easier to use the fact that momentum is never created or destroyed than it is to get involved with Newton's laws of motion, and all those tricky forces and accelerations.

## Using your momentum

As you noticed when you were diligently rolling that thuggish red ball into poor Jen Yu blue, it was difficult to get the collision quite straight, so that red stopped and blue escaped.

Often, red, coming in left to right, would kiss blue at an angle—and our nice little experiment would be ruined. Instead of red stopping, it would keep going one way, maybe slightly up the table as well as left to right, with blue going off at different angle, slightly down, as well as left to right.

When you think about it carefully, what's happening is quite simple.* After all, even average pool players know this behavior well, and of course it is all another result of the conservation of momentum. Let's look at this more closely—but first, time for a bet.

If you can gently mark on the table a straight left-right line that red is rolling along, I bet you that, whatever distance from that line (measured at a right angle from it) blue finishes, red will finish the same distance away, but on the opposite side.

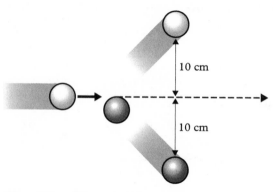

When balls collide at an angle

Tried it? Was I right? Of course I was. And you owe me ten bucks.

Anyway. Back to the physics. So although both balls go off at different angles, and they travel different distances left to right, at different speeds, if you tell me that

*Good pool players, however, do make the physics complicated. They also add in spin: partly to show how brilliant the human brain is at physics that is much too difficult for this book, and partly because that way they can get the cue ball into positions for their next shot that straight momentum would never allow.

red finishes 4 in. (10 cm) above the line, then I know that blue finishes 4 in. below, and all without looking or measuring.

Why? Well, as you've guessed, it's because momentum is conserved—and, more subtly, because momentum has direction, as we've just seen, and so it is conserved *in all directions*.

When just red was rolling, it had left-to-right momentum, and nothing up or down. This means that after the collision, our pool ball system, taken as a whole, still only had left-to-right momentum because momentum can't be created or destroyed. And therefore, after the kiss, the up momentum of blue must have been cancelled out by the down momentum of red, while the overall left-right momentum that was introduced by red was later shared between red and blue.

This momentum then gave the balls equal speeds *in the up-down direction* (because they have equal masses), and that speed was lost by both (to friction) at the same rate, and so in the up-down direction, they both travelled the same distance. Good, huh?

"But hold on a moment," I hear you say. "Those balls were rolling merrily along and they stopped. And something that's not moving has no momentum. So how on earth is momentum conserved when it has obviously been all used up?"

## You can't lose momentum

The clue, in fact, is in the "how on Earth?"

In the last chapter, we learned—through those risky driving experiments—that as objects slow and stop, energy is being conserved, but through friction, kinetic energy is converted into heat energy (which is in some ways a kind of kinetic energy, as we'll see later, because it is created by the vibration of atoms). So playing pool actually raises the temperature of that pool table.

Momentum, too, is always conserved. No matter how things bang into each other, or rub up against each other, or are magnetically pulled together, the overall momentum of the system is unchanged. But momentum can't change into anything else, like energy can: It's always there, even if it transferred to something so big (like the planet Earth, for example) that we don't notice its momentum change.

The key to the difference between energy and momentum is once again that momentum is a vector: It has direction as well as magnitude. The vibration of atoms doesn't give an object momentum, because those atoms are vibrating randomly in different directions.

Remember how, at the end of chapter 2, we noted that the car's braking affected the spinning of the world, because of Newton's third law? Well, the same law applies as our balls stop. The balls are stopped from moving by the friction from the pool table's felt, and the equal and opposite force from the balls pushes the table, and hence the floor, and the country and eventually the whole world!

This is great news for hippies, who get some justification for their belief that everything is connected, man. But it's bad news for them, too, because the effect of a pool ball on the movement of the Earth is so tiny that it is impossible to measure.

In short, hippies can't teach us much about physics. But perhaps Clint Eastwood can.

## The good, the bad, and the impossible

Just in case you're ever facing a six-gun-wielding outlaw in a graveyard where one of the coffins is packed with cash, it's worth looking at *The Good, the Bad and the Ugly* to find out exactly what won't happen.

First up, it is almost impossible to quickly draw a 150-year-old handgun and hit a person standing 55 yd. (50 m) away with a single shot. Heck, with those old guns you'd be pretty lucky to hit a barn door at that distance. But then, even if you got million-to-one-chance lucky, and did hit the bad guy, it is then completely impossible for the bullet you fired to knock that bad guy over backwards even though the recoil of your gun, which you're shooting from hip height, affects you not one bit as you stand there, cool as Clint.

After all, momentum has to come from somewhere. When your gun fires, and the bullet is blasted down the muzzle, its kinetic energy has come from a small explosion inside the gun. But momentum, as we've seen, can come only from other objects. So what gives your bullet momentum?

Initially, Clint and the bullet in his gun had no momentum. So, after he fired, they can't have had any overall momentum, either. So, whatever momentum the bullet had right after it was fired, Clint (the firer), should have had the same amount, but in the opposite direction. So, since the bullet hit Lee hard enough to knock him over backward, the recoil should have knocked Clint back, too, even though he was braced for the bang.

## It's been around and it'll keep going around

So far, we've been looking only at one kind of momentum, which physicists call *linear momentum* because it's all about things moving in straight lines. But now it's time for a quick spin around the other kind of momentum: *angular momentum*. It had a large say in the creation of this planet and its rotation. And, because it keeps gyroscopes standing upright, it is also a vital ingredient in the artificial horizon instruments that tell airline pilots when they're flying straight and level—and when they are diving at full speed into the ground.

Like its linear cousin, angular momentum is never created and never destroyed. Once there is spin in a system, it's there forever. In this way, angular momentum is a bit like politics: The spin can be transferred from one object to another and it may be not be noticeable for a while, but it never goes away.

Angular momentum is also important in those slightly dodgy Olympic events, such as gymnastics, diving and prancing on ice, where you get medals for looking nice. Platform divers, for example, have a deep-seated understanding of the fact that, as they spread themselves out, they spin more slowly, whereas as they tuck themselves in, they spin more quickly.

## Defying gravity

Let's try another experiment. First, take a wheel off your bike. Now, hold the wheel with two hands (one on each end of the axle), get it spinning quickly while upright, and then let go with one hand.

Come on, trust the physics: let go!

Amazing isn't it? Instead of falling over, your wheel is defying gravity, its angular momentum keeping it upright as you support it one-handed, on one side of the axle. Now you know why cyclists wobble like idiots as they start pedalling away from the lights but are far better balanced once their wheels are spinning quickly: it's all because the spin of their wheels keeps them upright.

Actually, your bike wheel's behavior is even weirder than it seems at first. After a moment, you'll realize that instead of falling over sideways, it wants to rotate around the point where you are supporting it. This behavior is even clearer if you support the wheel one on one side of the axle by hanging it on a piece of string. Then when you get it spinning, it will hang there, defying gravity, spinning, but also rotating gently around the string.

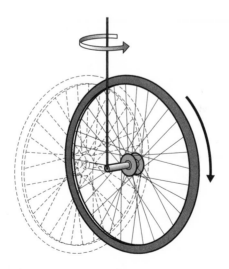

The reason? Well, instead of flat-out defying gravity, the wheel's spin subverts it.

## Gyroscopes

The weird spinning-bicycle-wheel-on-a-string contraption you've just been playing with is a crude gyroscope. The great thing about gyroscopes is that the spinning wheel inside them means they're not affected if you try to rotate them. So a gyroscope will balance on the point of a needle, or, more usefully, remain level as you fly your aircraft up and down, and bank left and right.

Combine a few gyroscopes with some acceleration sensors (that measure precession), and you can build a navigation system that helps keep careful track of your accelerations and rotations. And, if you know where you started, that information will always be enough to tell you where you are and how fast you're going. Systems like this are used in aircraft automatic pilots and, more destructively, in long-range missile guidance systems.

## A funny turn

Consider a point at the top of your bike wheel. If the wheel isn't spinning, the falling-over turning force applied by gravity and the string makes that top point fall down in an arc to one side. (Physicists and others like to call turning force "torque." More about that in a moment.)

If the wheel is spinning so that the top of it is moving away from you, there are no new forces involved, but the spin is already carrying our point at the top of the wheel downward, except that it is downward and away from you, rather than downward to the left.

Yet because the forces on our top point are the same as they were when the wheel wasn't spinning, the laws of physics say it has to get to the same place it would have if it wasn't spinning. And this is what happens: The spin of the wheel, added to the strange rotation around the string, gets our point on the wheel exactly where it should be, without the wheel falling—and so it doesn't fall.

Effectively, the wheel's spin turns the turning force caused by gravity through a right angle, so that it forces the wheel to rotate rather than fall. Physicists call this phenomenon *precession*: It's not a particularly useful term to remember unless you need an excuse for spelling procession wrong.

## Torque isn't cheap

You may have noticed that automotive journalists love the sound of their own voices. But besides talk, they also love torque.

Torque, as we mentioned, is turning force. It's what causes changes in angular momentum, making an object spin more quickly or more slowly. It's also a measurement of a car engine's ability to turn the drive shaft. So if the supercar you're testing has an engine that produces a lot of torque, the car will be fast and powerful. But the relationship between power and torque isn't straightforward.

Car engines produce different amounts of power as they run at different speeds. They are generally at their most powerful when they are running quite quickly—maybe at around 6,000 revolutions per minute for a typical gasoline engine.

But what automotive journalists love is low-end grunt. Grunt comes from high torque at low revs: It's the turning force that the supercar's thirsty V8 engine produces as it spins quite slowly; it's what starts the wheels turning; it's what accelerates all that shining metalwork and those swooping lines rapidly and exhilaratingly away from a standing start.

Lots of power at high revs helps you scream down the highway at more than double the national speed limit; but torque at low revs gets you from standstill to speed limit in 4.5 seconds, your heart pounding as you're pushed back into your seat by the acceleration, a mighty grin on your face as you conquer time and space and creeping middle age.

## Yu again

Before we're done with momentum, it is worth admitting that—despite what we found with our experiments in the pool hall—there *is* a way Jen Yu could have repulsed her attacker, much as she did in *Crouching Tiger, Hidden Dragon*, without breaking the laws of physics.

As Newton's first law showed us, the universe is lazy. Objects like to continue exactly as they are unless they get a proper kick in the backside from a resultant force.

To go back to our pool table for a moment, if we roll a Ping Pong ball into blue Jen Yu, she'll hardly move, but the Ping Pong ball will bounce straight off her. And don't think that this is because of friction. If we'd taken our blue ball up into orbit, set her floating in the middle of the room, and gently thrown our Ping Pong thug at her, the result would be exactly the same.

This little experiment shows us two things: first, don't use a Ping Pong ball as the cue ball if you want a decent game of pool; second, that it isn't only momentum that is important.

## The lazy universe

We can say more about this tendency of the universe to stay as it is. We know that the bigger and heavier an object is, the more difficult it is to get it to change its movement (or get it moving if it isn't moving in the first place). An object that isn't moving doesn't have any speed/velocity, so it doesn't have any momentum. But it does like to remain at rest. And the heavier it is, the lazier it is.

Physicists call this resistance to change in motion "inertia," which can be a little confusing for the rest of us who tend to equate the inertia of a moving body with its momentum. But it is worth separating momentum and inertia.

Heavy things have great inertia: Whether they are moving quickly or slowly, or not moving at all, it is difficult to alter their motion (go push an oil tanker, if you don't believe me).

And light things can still acquire great momentum if they are moving very quickly. The Man With No Name's bullet (if he were a real nameless man, in the real world, firing real bullets, etc.) would weigh only 0.1 oz. (16 g). Throwing that at a bad guy wouldn't have much effect, but firing it at him at around 300 m/s would give him quite a jolt.

It turns out that we measure inertia in kilograms. An object's mass (which, on Earth, because of gravity, gives it weight) also gives it inertia. And for many things that are impossible to weigh (you can't, for example, put a planet on your bathroom scales), a calculation of their momentum—and therefore their inertia—gives us their mass—the measure of just how much stuff they have in them.

We'll see more about the stuff that makes mass in just a moment. After we've tied up one last loose end...

## Secret weight problem

So was that *Crouching Tiger* scene realistic, after all?

Jen Yu may have looked like a svelte young kung fu fighter, but did she in fact she have a weight problem? We assumed in our earlier pool table experiments that Jen's attacker weighed about the same as her. But if instead she weighed much, much more than him, then our new experiment, with the thug played by a Ping Pong ball, would be a better indicator of the truth of that scene.

So, maybe Jen had massive weight problem. The kind of massive weight problem where she was something like a couple of tons overweight. Then even a heavyweight thug might bounce off her—if she was made out of something bouncy, and as long as her massive weight problem hadn't already caused her to crash through the floor of that rickety restaurant where she was sitting upstairs eating before she (and her attackers) kicked it all to pieces during their fight.

But how did she get to weigh that much eating low-fat rice and noodle dishes and drinking green tea? And how, if she weighed a couple of tons, did she manage to leap about like a flea?

Oh well. My advice is, don't watch an action movie if you want to learn physics.

# 5

# Heat and Matter

*In this chapter we'll learn all about matter and what it's made of. And then we'll see how what-it's-made-of explains how matter can turn from a hard, springy solid into an airy-fairy gas; why water is weird; why smoke gets in your eyes; and how life on planet Earth keeps us all under pressure.*

Matter is the stuff all around us. It takes up space and it has mass: That's why it's filling up your cabinets and making your shelves creak under its weight (though of course those cabinets and shelves, and you and your house, and all the world are made up of matter, too).

Apart from having mass and volume, and creating storage headaches, what else does matter do? Maybe if we find some answers to that, we can find out a bit more about what matter is.

## What's the matter?

Thousands of years ago, ancient Greek and Indian philosophers came up with the idea that if you kept dividing matter up into smaller and smaller pieces, you'd eventually end up with something fundamental: something that just couldn't be split again. That something was called the *atom*.

But it was a long time before the new science of chemistry began to give these ideas some experimental basis. In 1803, an English scientist named John Dalton noticed that certain elements could be combined with others in a ratio of small whole numbers. So if you had a big, well-equipped chemistry lab and piped 250 gal. (1,000 l) of hydrogen and half as much oxygen into a container, then threw in a lighted match, there would be a deafening explosion and all the gas, once it had cooled, would become enough fresh water to fill a large glass.

(If that explosion has made the water hard to find amid the general wreckage of your lab, get a mop—we're going to need that glass of water for the next part.)

If you use alternative volumes of oxygen and hydrogen (for example, the same volume of each, or twice as much oxygen as hydrogen), there will always be some hydrogen or oxygen left over. That's because water is $H_2O$: water molecules—the smallest components of water—are made up of two hydrogen atoms and one oxygen atom. So you need twice as much hydrogen as oxygen to make water.

## Chemistry is elementary

Today, science has discovered 117 chemical elements, although only 94 occur naturally on Earth. You know many of them: hydrogen, helium, carbon, oxygen, iron, copper, silver, gold, chlorine, sodium, and neon are a few of the most familiar.

In chemistry, the atoms that make up these elements are basically as small as you can get. But they are rarely found sitting around by themselves—they're too sociable for that. For example, almost all the hydrogen on Earth is in water molecules because, as we just discovered, it takes very little persuasion to mix it up with oxygen and make up a cozy little threesome.

Atoms are so tiny that it is impossible to see them, even with a microscope. A pinhead is around 4,000,000 atoms wide. (Except that physicists have found that, like clouds, atoms don't really have fixed boundaries, so measuring them is difficult.)

Physicists have also found out how to trump the chemists' work by smashing the atom to make even smaller bits of matter. But let's not get ahead of ourselves. There's lots to find out about atoms before we start breaking them.

## Ice breaker

Now we've made a glass of water, what else can we find out about it? Well, the obvious thing is that if we make it very cold, our water will become ice. If we then warm it, it will become water again.

In solid ice, our water molecules are pretty much frozen in place, packed tightly together and held by long-lasting bonds they form with their neighbors (although these bonds do allow them to vibrate around a central point).

As they warm up, the extra energy helps our water molecules vibrate more and more until they break free of these stable relationships to form a liquid. The molecules are still packed tightly, but their bonds with their fellows are fickle and fleeting as they jostle about like horny teenagers at dance club. Now they can flow over each other to fill whatever container they're in. (Note that this change of state—from solid to liquid—takes a lot of energy, which is why by melting, ice cubes can cool a drink much more than a similar amount of ice cold water could.)

Heat your water to boiling point (which takes lots more energy) and the molecules will gain enough oomph to escape the dance and become free spirits—in other words, they form a gas. In this state, our water molecules have broken all their bonds and are really speeding around, so they spread up and out in all directions until gravity hauls them back or they bump into something solid.

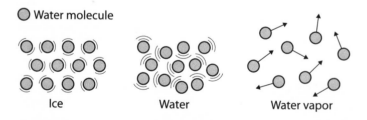

You can tell that gas molecules are much more spread out than those in solids and liquids because gases are relatively easy to compress (it only takes a bike pump). It is almost impossible to compress solids (unless, like sponges, they have a structure that's full of air) and liquids (which is why hydraulic systems work: try to compress oil at one end of a tube—with your car's brake pedal, for example—and it'll push back at the other—making your brake pads grip your brake discs).

## Steamed up

The gas formed from water is called water vapor. Even though it took a combined volume of 375 gal. (1,500 l) of hydrogen and oxygen to make our glass of water, when heated that water makes around 125 gal. (500 l) of water vapor at room temperature. This is because it's not the size of the molecules that determine the volume of a gas, it's purely just the number of them.

Water vapor is as clear and as colorless as air. The steam we see coming from that saucepan (in which we're boiling away our precious cup of water) is slightly

### States of matter

Solid, liquid, and gas are three states (although physicists like to call them "phases") of matter. But there are quite a few others, too, though they are much less common, and only created when there is enough energy around to break up the atoms themselves. They include plasma, Bose-Einstein condensates, and Fermionic condensates.

This knowledge is useful only if you feel like arguing with someone who tells you that there are three states of matter.

different: It, like the clouds we see in the sky, is actually made up of tiny drops of liquid water hanging in the air.

Soon all the water we made will be gone, boiled off and dispersed into the atmosphere. But we can bring some of it back. Water vapor will condense into water on a cold surface, the cold stealing away the energy of molecules that hit that surface, so that they can no longer fly free.

This is good news for us—water is amazing stuff and we're going to need it again in a minute.

## Water: strange thing

Almost all solids expand slightly as they get hotter. That's why large concrete structures have those rubbery expansion joints in them. And almost all solids expand a little further as they turn into liquids. But water is a little odd.

Ice does expand slightly as it is warmed, but then it goes the other way and contracts as it turns into water. The shape of water molecules, and the fact that one side of them has a slightly positive electric charge and the other a slightly negative one, means that they can fit together more closely when jostling around in liquid form than as a solid (in which all those positive and negative charges line up in an orderly fashion, like a grid of mini magnets).

So water expands as it freezes: bursting pipes, but also causing ice to float, which allows fish to survive beneath it. Water also has many other properties (such as surface tension) that are found in few other liquids, and as far as we know, it is uniquely suited and vital for life as we know it.

But for our experiment, water is still a good way to find out about solids, liquids, and gases. And after all the trouble we had making our glass of water, it would have been a shame not to use it.

## More about molecules

OK, it's not easy to believe that your still glass of water is full of jostling molecules. But if you mix a teaspoonful (5 ml) of tiny grains of pollen into your water and then look at them through a microscope, the evidence is right there before your eyes. You'll see those pollen grains jiggle as they're battered by the water molecules.

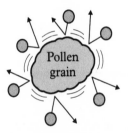

Brownian motion

This is called Brownian motion, after Robert Brown, the Scotsman who discovered it: You probably saw it at school. If our water is likened to a horny teenage dance, then the grain of pollen is like a giant balloon on the dance floor that is being continually bumped this way and that, but never very far, by the crowd of dancers around it.

In fact, molecules are 250,000 times smaller than pollen grains. We know this because Einstein did the math behind Brownian motion, which, if we measure our pollen grains, allows us to calculate the size of the water molecules. You can't see them, but they really are there.

Brownian motion can be seen in gases, too. Smoke particles in air (which is a mixture of 78 percent nitrogen and 21 percent oxygen, plus tiny amounts of argon, carbon dioxide, and water vapor) move like pollen grains in water. And that's just the start of the physical phenomena that can be explained once we realize that molecules and atoms are always moving.

A hot-air balloon fills up and then floats away because of a whole host of these effects. Come on, let's try it. You'll never have a better excuse to fly around in a wicker basket—with just the laws of physics for support and no control over where you go.

## A lot of hot air

We've already seen how hot-air balloons rise because they displace a greater mass of air than they contain. Now we can discover why this happens.

Time for a quick thought experiment. And thought only, please, because this one is dangerous.

Imagine a sealed box full of air. Inside, free-spirited molecules are shooting around in straight lines at an average speed of 500 m.p.h. (210 m/s) until they bump into each other or, as is more likely because they're well spread out, into the walls of the box. This steady drumming of molecules on the inside of the box is what we know as *pressure*. It's a bit like what we feel from a strong spray of water, except those free-spirited molecules push out in all directions.

Now, if you put that box over a lighted fire, the air molecules will get fired up. They'll move more quickly, so they'll hit the walls of the container harder (free spirits don't like being kept in boxes). The pressure on the walls of the box rises. Keep your box on the fire long enough, and bang: It explodes. Told you it was dangerous—you can't lock up a free spirit forever.

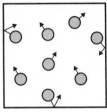
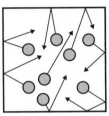

Low pressure          High pressure

## Feel the pressure

Pressure is measured in Pascals, and sometimes p.s.i. (pounds per square inch). It's a measure of the force acting on a given area: So if you halve the area, you double the pressure. This is how snowshoes work: By spreading your weight (and remember, weight is force) over a wide area, they reduce the pressure on the snow and prevent you from sinking into it. If instead you went walking on the snow in high heels or on stilts, their tiny surface area would concentrate your weight into lots of pressure on the snow, and you'd fall straight through it.

But we're all under pressure anyway: 62 miles (100 km) of atmosphere presses down on us with 100,000 Pa of pressure. This is why suction cups work: When you attach the cup to a surface, some of the air is squeezed out, lowering the pressure inside; now the higher atmospheric pressure is pushing the suction cup to the surface harder than the cup is pushing away from the surface.

## A bit about Blaise

The unit of pressure is called the Pascal, after French scientist and philosopher Blaise Pascal. Born in 1623, he did plenty of brilliant math and science from an early age, including building a mechanical calculator and essentially inventing probability theory. He was also instrumental in the invention of the hydraulic press, so it's appropriate that the unit of pressure carries his name.

After a vision in 1654 and an apparent miracle healing of his daughter in 1657, Pascal gave up all that science stuff and devoted his life to God, while writing arguments that reason was untrustworthy. He died in 1662, at age 39.

## Blowing up a balloon

Now we know a bit more about pressure, let's try that box-on-the-fire experiment again, except with some sort of imaginary fireproof party balloon. Make it a red one with yellow lettering that says, "Happy Birthday."

Once you've blown it up and tied the end, the pressure inside equals the air pressure outside—that's why it stays at the same size. Now stick this fireproof balloon on the fire for a little while. What happens? As the air inside grows hotter, our free-spirited molecules start hitting the inside of the balloon harder—the pressure rises and so the balloon gets bigger.

Now, as the balloon blows up, the free spirits have more room, so they don't hit the walls quite so often. Eventually—if you take the balloon off the fire so that the air inside reaches a constant temperature—the reduced number of hits on the walls compensate for the extra energy of those hits. Once again, the balloon will reach a constant size. It'll be bigger, but inside and outside pressure will again be equal. All the fired-upness of those free spirits will have bought them a little extra room.

Careful with that fireproof balloon, though. That expanded air will be lighter than the rest of the air, and so it will try to float away. Something similar happens to the hot air heated by our hot-air balloon's burner: It expands and rises immediately. This process is called *convection,* and it's a common way for heat to travel. Columns of warm, rising air above hot objects are useful for filling hot-air balloons, or keeping birds of prey gliding while they look for small furry snacks. But they can be a problem, too.

Imagine you're back from a freezing trip in hot-air balloon. (Did you know that the reason it's cold up high is that the air pressure is lower? Can you see why?) Now you're back inside and there's a nice hot radiator on the other side of the room. So why is there a cold draft on the back of your neck?

Well, the hot air above the radiator rises, and as it does, it slowly cools. More rising air pushes it away from the radiator and eventually, when it's above you, it has cooled enough to fall—right down the back of your neck. Ugh!

But don't moan. You're not really cold. In a moment, well find out what it means to be so cold that you can't get any colder.


## Absolutely freezing

Now we know that temperature is really just a measure of how much the molecules in matter are bumping around, we can think about what might happen if they stopped moving completely.

Well, we can think about it, but it can't happen: The third law of thermodynamics says it is impossible for matter to lose absolutely all of its thermal energy and so reach the complete lack of temperature that physicists call "absolute zero."

The strange thing about absolute zero is that it's only −460°F, or −273.15°C (and water freezes at 32°F/0°C). That is extremely cold; but when you consider that +523°F (+273°C) is a good temperature for cooking thin, crispy pizzas, and the middle of the sun is estimated to be at 27,000,000 °F (15,000,000°C), it doesn't seem that cold.

Then again, scientists do have ways of cooling things to temperatures very close to absolute zero, and it is in fact so cold that strange things happen—like, for example, the formation of those Bose-Einstein condensates we mentioned earlier.

## Why springs are springy

Before we finish with the hot and heavy stuff and find out what makes light work, we should quickly mention a couple of the many other familiar things that are explained by the kinetic theory of matter (which just means that the molecules jiggle around as we've described). Let's look at why puddles evaporate without boiling and why springs are springy.

Puddles disappear even on cold days because it's the average energy of the molecules that give the puddle its low temperature. But that average disguises the fact that, at any time, the molecules are actually moving at many different speeds through their packed teenage dance club. And the ones that are lucky enough to get enough energy when they're near the surface of the water can—if they get a straight run—fly free, up into the sky, breaking their watery bonds and leaving behind their teenage angst (at least until they turn into rain or condensation). And so the puddle slowly evaporates.

Springs are springy because the atoms in a metal are all bonded, like the molecules in ice. Stretch them apart and the bonds pull back. (The molecules in ice are less tightly bonded and the pattern of their bonds is different from that of the atoms in a metal, so if you pull ice, it just breaks apart: it doesn't make a good spring.)

## A small mistake

In this chapter we've been thinking of matter as made up of little bouncy balls. Well, it's not.

Don't dismiss this idea, though. It's what Newton believed. And, as we've just seen, it's a very useful way to think about all kinds of physical phenomena.

It's just that we have plenty of evidence that although atoms and molecules sometimes behave like bouncy balls, they are actually made up of lots more tinier bits, and that these bits are so tiny and so different from the matter cluttering up our world that almost everything about them seems strange and wrong.

And to start finding out about the world we can't get our hands on, we need to take a trip to the shore.

# 6

# Waves

*In this chapter, we'll see that all waves are similar: They travel through each other unharmed, reflect when they bump into some things, and bend as they enter others. And yet waves take many different forms. So we'll find out why wave power is tricky to harness, we'll calculate the speed of sound, and we'll hold proof in the palms of our hands that light is a wave.*

**B**ut let's start at the seashore, because it is a great place to find out about the physics of waves. That's because—as you'll have guessed already—the sunshine, the waves on the sea, and the sounds of the seaside are all waves. So come on, grab your surfboard and we'll swim out to discover more.

Once you're through the surf, and sitting waiting for a nice big wave to ride back in on, you'll be noticing the most important feature of water waves: that they lift you up and down.

And there's another important feature of waves that's obvious, too: The waves carry energy to the beach (which is why they can push you along when you find one to ride), but the sea stays where it is. Sure, the surf plops up the sand a bit, but the movement of the waves and the energy they carry (toward the shore) is independent of the movement of any bit of the sea (which basically goes up and down with a bit of backward and forward so that each point in the sea actually goes round and round in a circle).

So there you go. You haven't even had to show off your surfing skills and you've discovered what a wave is: a transfer of energy without a transfer of matter.

But don't ride in yet. There's lots more we can discover out here about water waves, and—because all kinds of waves do the same wavy tricks—about other kinds of waves, too.

## Interfering busybodies

For example, while you're out at sea moving gently up and down, you can make your own little waves by dipping your hand in the water. These ripples spread out on the sea on top of the bigger waves that you aren't riding. So it's clear that lots of waves can travel through the same stuff, in lots of directions, and all at the same time. They don't bump into each other like balls.

But they do interact in another way. Your ripple makes the big waves a little higher in some places, and a little lower in others. Ripples on a big swell are hardly notice-able, but if you drop a couple of similarly sized surfers into some reasonably calm water, it's clear that the rip-ples make a new pattern as they spread out, because in some places you get two troughs together (and the result is a doubly deep trough) and in others you get two peaks (and the result is a doubly high peak).

## Sea power

Waves that are 3 ft. (1 m) high hit every 3-ft.-wide section of beach with enough power to run 10 strip heaters—which is why it is frustrating that we haven't been able to capture wave energy to generate electricity.

Part of the problem with wave power is the way waves carry it. The energy of a wave is proportional to the square of its height. So a 9-ft.-high wave has 3 x 3 = 9 times more energy than a 3-ft. wave. And when a hurricane brings 90-ft. waves, with 900 times the energy of the little waves you were generating power from, your clever wave power gadget is turned into something mangled and useless.

In yet other places a peak in one surfer's ripples meets a trough coming from the other surfer, and then at that point the waves cancel each other out. It seems as if there is no wave there—but of course the wave energy from both surfers is still travelling through that point.

Physicists call this interaction of two or more sets of waves *interference*. The poor surfers caught up in such wave interaction might have a slightly stronger word for it.

Waves from first surfer

Waves from second surfer

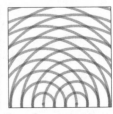

Interference pattern from the two sets of waves

## Breaking waves

Waves travel at different speeds in different things. And shallow water is a different thing from deep water. Waves notice this difference, because as the water gets shallower they start scraping along on the sand.

This slows them down, but the slowing is more pronounced at the bottom where the sand is, and has less and less effect the farther up the wave you look. So the top of the wave carries on at its same old speed, and so the top starts to move ahead of the rest of the wave below, and eventually it moves so far ahead it falls over, and the wave breaks.

## Turning a corner

One result of a wave moving from the deep sea, where it moves at one speed, to a shallow bay, where it moves more slowly, is that so long as it hits the boundary of deep and shallow at an angle, it will turn a corner.

This makes sense if you think of our breaking waves. Just as the top wasn't slowed as much as the bottom and so the waves broke, if part of a wave has reached shallower water, that part will slow down, whereas the part that is in deeper water will be moving faster, and so the wave will change direction. This is called *refraction* of the wave.

This situation is a bit like the way in which a bulldozer steers by making the caterpillar tracks on one side move more quickly than those on the other. If its right track turns more quickly than the left, that side of the bulldozer moves forward farther in any given amount of time than the left, and so the bulldozer turns to the left.

## Mind the gap

We've already found one way to get waves off the straight and narrow, but there are others, too.

On a stormy day, with huge waves rolling in from the ocean, it's clear that the waves make it in through the small gap in the harbor wall and spread out, gently rocking all the boats as they sit at their moorings.

One reason for this is *reflection*. When wave energy hits a hard wall, it has to bounce off; so the wave is reflected and travels back through itself. Throw another surfer into a calm harbor and you'll see the ripples he makes bounce off the walls and spread out.

But if the waves are coming straight through a reasonably narrow gap in a harbor wall, you can see another wave effect: *diffraction*. Instead of just the portion of the wave that hits the gap continuing on inside the harbor while the rest of the water remains calm, the wave spreads out a little beyond the wall, and that spread grows the farther

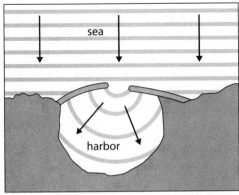

Waves diffracting in a harbor

the wave travels from the gap. The smaller the gap, the greater the effect.

## Transverse and longitudinal waves

Water waves are *transverse* waves. They involve something moving up and down as the wave goes along. If you look at them from the side, they make that familiar wavy line, a bit like a long train of camels.

*Longitudinal* waves, on the other hand, involve something moving backward and forward in the same direction as the wave travels. But just as with transverse waves, the matter itself doesn't go far: just back and forth a little as the wave travels on.

Although these two kinds of waves might seem quite different, they do both display all the strange wavy properties we've been looking at. Both kinds of wave will reflect, refract, diffract, and interfere. And it's not just water waves that do this.

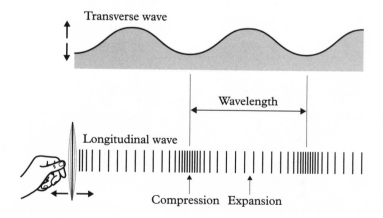

## Sound thinking

Sound is a great example of a longitudinal wave. You can picture what's happening if you pluck a note on your guitar. OK? Ding! See, your G-string is vibrating: moving quickly toward you and away again. And as it comes toward you, it squeezes up the air molecules, creating a "compression." Then, as it moves away, it squeezes the air molecules on the other side, leaving the air on your side slightly spread out (and so at slightly lower pressure—a "rarefaction").

Then it comes back and you get another area of squeezed-up (or slightly higher-pressure) air traveling your way as the laws of conservation of energy and momentum ensure that the movement of one molecule is transferred to the next.

## The Doppler effect

We've all heard the way ambulance and police sirens sound different coming and going. It's because of the *Doppler effect*: a phenomenon named after the Austrian physicist who identified it, long before sirens were even invented.

Think of a typical teenager's car, the sort that's as much a boombox on wheels as it is a mode of transport. And let's imagine that, every second, it throws one pulse of high-powered bass in your direction as you sit there on the beach muttering to yourself about kids these days.

Now, let's say our boombox-car is moving so quickly toward you that every successive pulse is released at a distance so much closer to your ear than the previous one

## A tuneful note

Even though you may not be the best guitar player in the world, the regular vibration of that G-string, back and forth 196 times a second, means it makes a nice musical note.

Busy physicists don't have time to say "times a second" every time they talk about how fast a guitar string is vibrating or a wave is going through its full cycle from peak back to peak, or compression to compression; so instead they say talk about *frequency,* which they measure in *Hertz* (written as *Hz*).

So the open G-string on a guitar has a frequency of 196 Hz; those surfers still waiting out at sea have a frequency of 0.1 Hz (that is, every 10 seconds they are at the peak of another wave).

Now, if you press down on that G-string halfway down the neck, it will vibrate twice as fast when plucked—moving back and forth 196 x 2 = 392 times per second. Musicians will be able to tell that this new note is still G, but a different G, one an octave (that's eight full notes of the major scale) higher.

So that's math and music in perfect harmony. If you halve the length of the string, you double the frequency of the sound.

that it will spend half a second less traveling that ever-decreasing distance to you than the previous one.

Can you hear what's happening? Instead of your ear picking up a pulse every second, it receives one every half a second. That's twice as many pulses per second, which

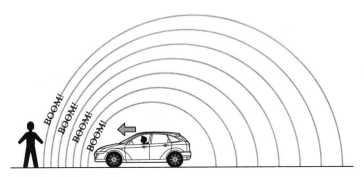

Doppler effect

means you hear that constant bass rhythm pumping twice as fast as the driver does (and the note will sound higher-pitched). Likewise, after it drives past, you'll hear the rhythm slow to half of what the driver is hearing (and the pitch of the sound will be even lower than what's rattling his fillings).

There is one important footnote to the Doppler effect: What we hear depends on the fact that the speed of sound in air is always the same.

Sound waves from something coming quickly toward you don't travel any faster than those from something that isn't moving. They're not like balls thrown forward from a moving car. This means the speed of sound in air is a *physical constant*.

## Enjoying the rays

Light travels at 299,792,458 m/s in a vacuum. That's really quick. It's also a universal constant. It's always that fast, no matter where you are. That's why the length of a meter is

## Electromagnetic waves

After it was discovered that electricity and magnetism were related (more about this in chapter 7), in the 1860s a Scottish mathematician named James Clerk Maxwell created and published a set of equations, known today as Maxwell's equations. These equations showed how electric charges create electric fields, and how electric currents create magnetism. Together they showed how electric and magnetic fields travel through space in the form of waves, and how they move at the speed of light (and how fast that speed is).

This was pretty good stuff, but best of all they showed that light is purely a combination of electricity and magnetism. That's why physicists call light an *electromagnetic wave*.

To us, it is special because we can see it. But it is different from many other types of electromagnetic wave only in its wavelength and frequency. The radio waves that bring us music and chattering DJs, the microwaves bounced around in microwave ovens to heat food, warming infrared,

based on it.* It's also, as we'll discover in a few chapters' time, one reason Einstein discovered special relativity.

For the moment though, we'll note that all this rushing around makes it much harder to get a handle on light

*In the 18th century the meter was defined as one ten-millionth of the distance from the equator to the north pole. Which is nice and neat, though unfortunately it was later discovered the calculation was one-fifth of a millimeter wrong. So it now has to settle for a less handy definition: one meter is the distance light travels in a vacuum in 1 divided by 299,792,458 seconds.

ultraviolet (which tans our skin), X-rays (which enable us to see the bones beneath our flesh), and dangerous gamma rays are all the same kind of thing. They're all electromagnetic waves like light.

Long-wave radio starts at a frequency of around 10,000 cycles per second (or Hertz) where the waves are a kilometer long (0.6 mile); whereas high-energy gamma rays go up beyond $10^{24}$ Hertz (with a tiny wavelength of $10^{-16}$ m). Somewhere in the middle are the frequencies that we see as different colors. Red light has a frequency of around 4.4 x $10^{14}$ Hz; blue a frequency of around 6.4 x $10^{14}$ Hz.

The more you think about this, the more amazing it is. It's worth repeating: All these invisible forms of electromagnetic radiation—which is created by the stars and heaters and radio transmitters—are basically the same kind of thing as visible light (better known to most of us as "light" because obviously, invisible light isn't light. It might cook your dinner, but it's not going to help you see in the dark).

than on sound, which is one reason physicists had a long argument about whether light was in fact made up of waves or particles.

## In the dark

The Greeks started the confusion. They thought something came out of our eyes and lit up what we looked at. Explaining why we couldn't see in the dark was always tough for the Greeks.

In 1670, Newton said light was made from particles (or, as he called them, *corpuscles*). That made it easy to explain reflection and the fact that light travels in straight lines and through vacuums (especially for Newton, who, as we've seen, was very good at the physics of motion). Light could now travel from the sun, bounce off things around us and then enter our eyes. In the dark, there was no light, and so nothing reaching our eyes. But Newton did find it tricky to account for refraction and those other weird wavy properties.

Other scientists, including the Frenchman René Descartes (of "I think, therefore I am" fame) and the Dutchman Christian Huygens believed that, because it could be diffracted and refracted, light was more likely to be made up of waves.

The debate raged for over a century, until a bright Englishman named Thomas Young, who could speak fourteen languages by the time he was 14, came up with the not very elegantly titled *double slit* experiment.

## Fringe benefits

What Young did was to use light waves of a single color, and *two* very narrow slits, 0.004 in. (0.1 mm) apart, for the waves to diffract through, instead of a single gap. (The thinness and closeness of the slits is important because visible light has a wavelength of around 0.0005 mm.)

The result was that, because there were two slits for the light to come through, it was as if Young had created two separate but identical sources of light. And when the light waves spread out from these slits by diffraction, they interfered just like water waves, creating some areas where the waves reinforced each other, and others where they cancelled each other out.

Best of all, this interference could easily be seen with the naked eye. In a dark room, Young's slits make a pattern on a screen: a central bright spot, with successive fainter blobs of brightness (called *fringes*) slowly fading out on each side. Physicists can even use the spread of this pattern to calculate the wavelength of light.

So there you have it: one light experiment, two wave-related effects. Particles can't diffract or interfere, so poor old Newton's theory was blown out of the water.

If he hadn't been dead, he'd have been furious!

## Spectacular light show

Here on the beach, the fact that light is a wave is obvious.

Let's start with those binoculars that you're using to study the fine array of surfing talent sitting out there on the waves. These use lenses. And lenses make things seem bigger by refracting the light so it bends, because light travels more slowly in glass than in air.

Also, can you see that spray bow above the waves? (It's that rainbow formed in the sea spray when the sun is behind you.) It's there because the white light coming from the sun is really made up of many colors. And light of different colors is refracted different amounts (because it has different wavelengths). So when white light goes into a drop of water (and refracts), reflects off the back of the drop, and then comes out the front (and refracts again), this double refraction causes it to be split so much that a rainbow appears.

## Can I be invisible?

In theory, if you bend light around something and then straighten it out again afterwards, that something would become invisible. That something would be a bit like a rock in a stream: a little way downstream of the rock, the water has settled down, and there is no way to tell (by looking at the stream) that it ever passed a rock.

Theory, of course, is one thing; practice, quite another. But in 2006 and 2009, teams of scientist in the United States and the United Kingdom made newspaper headlines when they claimed that they really had made objects invisible using light-bending cloaks made of "metamaterials"—substances engineered to have a precise structure, such as tiny hairs or holes, on a scale similar to the wavelength of light.

However, the first "cloak" was actually a heavy and completely-not-invisible metal cylinder that only bent microwaves. The second was almost invisible—but only because it was a hundredth of a millimeter wide. And again, it had no effect on visible light. It only worked on infrared. Many other scientists pointed out that it is difficult to see how these "cloaks" could ever work for more than a few very specific wavelengths (and therefore colors) of light. The metamaterial would have to be very ingenious; otherwise, like raindrops, it would split white light into a spectrum as it bent it, and you would end up hiding under a rainbow.

So, for the foreseeable future, invisibility is out of reach for us mere mortals (though as 90 percent of the universe is invisible anyway—see chapter 10—perhaps that doesn't matter).

The color of the sky is also a clue to light's waviness. The sky is blue because blue light, which has a shorter wavelength than red and yellow light, is, because of that wavelength, scattered by the air molecules bouncing around in the sky. So whichever direction you look (as long as it's up), some of this scattered blue light reaches your eyes, and the sky looks blue. Most of the red and yellow light travels directly to the surface of the Earth and is reflected and absorbed by the sand, the sea and, of course, that surfing talent.

The clincher for light being made of waves is this: You can do your own single-slit version of the double-slit experiment. But we'll get to that in a moment, after a quick look at how the waviness of light could help us all disappear.

## Two fingers to light particles

Young used two slits because it clarifies the interference caused by diffraction. But you still get an interference pattern with a single slit because the light from different sides of the slit travels different distances and still arrives in or out of sync. So here's the handy version of the experiment.

Ready? Hold two fingers very close together so there's a tiny gap between them that you can barely see through. Now look toward something bright, like that white sea wall (not the sun). And there (by the laws of physics and the fact that light is a wave), in the gap between your fingers and running parallel to them, will appear faint dark lines of destructive interference ("dark fringes"). You can see they aren't due to blurred vision if you move your fingers apart a bit farther and watch them disappear.

Isn't that amazing? And isn't it even more amazing that Newton's theory of light being made up of particles was believed for a hundred years?

Well, yes, it is. But there is one thing even more amazing than that: In chapter 9 we will discover that Newton was, in fact, right. Light might behave as a wave, but it *is* made up of particles; only now we give them a fancy name. We don't call them corpuscles (which sound like the sort of nasty blisters that might have afflicted medieval alchemists). We call them *photons*.

# 7

# Electricity

*In this chapter we'll learn about static electricity, batteries, voltage and current, Ohm's law, the relationship between electricity and magnetism, and DC and AC current. Put them together and you get (a) the building blocks of our powered-up, gadget-crazy world and (b) some really dangerous physics.*

Today, one form of electricity is regularly rediscovered by shoppers in shopping malls, who get a shock from touching a metal elevator button after a couple of hours shuffling around on hard-wearing carpets (though it's a mild shock compared to the one they receive when they open their credit card bills a few weeks later).

Electricity of all kinds is essentially the movement of *electric charge*. And electric charge is created by something having more or fewer tiny negatively charged particles, called electrons, than it should. (We'll talk more about electrons later.)

Charge builds up because different materials behave in different ways toward electrons. Some, like certain

## Franklin's Kite Experiments

When Benjamin Franklin wasn't plotting with the French to end British rule and found the United States, or working to abolish slavery, it is said that he liked nothing better than a bit of extremely dangerous practical physics, such as flying a kite into thunderclouds to collect electric charge.

This charge was conducted to the ground along the kite's dampened string, and the tiny sparks that jumped from a key tied at the end showed that this electricity was the same stuff as the static electricity that had been discovered long ago by the ancient Greeks, who found that their amber jewelry attracted dust because it picked up a charge after rubbing on their clothes.*

Franklin's kite demonstration is so risky that we won't even try it as a thought experiment. But Franklin was aware of the danger, and he certainly didn't try it when lightning was flashing. In fact, it's possible that he never did the experiment at all, and simply wrote it down as a good idea. Others, including a physicist named Georg Richmann, proved that in fact it wasn't a good idea. Richmann was killed by lightning during a similar experiment.

carpets, are inclined to collect extra electrons if they get the chance, and so become negatively charged. Others, like leather shoes and human skin, are happy to give away electrons and become positively charged.

---

*In fact, the word electricity comes from the Greek work elektron, meaning amber.

The electrical charges that result from this trading of electrons are static electricity—electricity that isn't flowing. But the charge does flow when the unfortunate shopper touches an electrical conductor, such as that metal elevator button. This brief flow of charge is called *electrostatic discharge*—the discharge of static electricity. This is what gives our shopper the (small!) electric shock.

Electric shocks through metal fixtures might be an annoyance to bargain hunters, but sometimes it can be very useful. Benjamin Franklin's knowledge of electricity led him to invent the lightning rod: a metal strip that could take the electrical charge from lightning straight into the ground, diverting it before it entered and wrecked tall buildings with fires and explosions.

At the time, churches, as the tallest buildings in town, were often struck by lightning. In fact, at one time people were advised to escape from thunderstorms by going anywhere *except* a church. Today lightning rods are a common sight on tall buildings, and ships and aircraft have lightning protection that works using the same principle.

## Staying positive

It wasn't until 1896 that electrons were discovered, by the British physicist J. J. Thomson. But for a long time it had been known that if you used a powerful battery to give a strong negative charge to a piece of metal at one end of a tube containing nothing but a vacuum (that is, containing nothing at all), and a strong positive charge to a similar piece of metal at the other end of the tube, rays of something mysterious would shoot from the negative cathode to the positive anode.

## When life gives you lemons, make batteries

It's quite easy to make a simple battery cell with some ordinary household objects. Take a lemon and stick a piece of zinc in near one end (we'll use a galvanized nail). Stick a piece of copper (like a copper coin) in near the other end. Make sure both pieces go into the flesh of the lemon but don't touch each other.

The two metal parts are called *electrodes*—they're the parts which electric current will flow from or to. The zinc (nail) electrode is the *anode,* or negative terminal, while the copper electrode is the *cathode,* or positive terminal. In a circuit, electrons will flow from the negative anode to the positive cathode.

The lemon battery works by a chemical reaction—the acid in the lemon dissolves the zinc, releasing electrons at the anode and thereby making it negatively charged,

Thomson showed that these rays were made up of tiny particles 1,000 times lighter than a hydrogen atom, and that they were the same no matter what substance you made the cathode from. He also discovered that these same particles could be produced by shining the right kind of light on certain metals and by radioactivity. They came to be called electrons.

But plenty of inventors hadn't waited for the physicists to "discover" what they had been playing with for years. The first electric light was demonstrated in 1803 by British inventor Sir Humphrey Davy, and by 1880 Thomas Edison started selling practical lightbulbs in the

Lemon

while a copper-acid reaction at the cathode absorbs electrons, giving it a positive charge. Unfortunately this battery is very weak (the lemon's citric acid isn't very strong, after all)—so weak, in fact, that to see it working you'll need to either use an electronic multimeter, or make several lemon batteries and use them all at once to power something. But be warned—you'll need nearly 10,000 to power a flashlight bulb.

United States. Soon electric gadgets were helping us get around, get fed, clothed, and watered—and get out the message about just how many new kinds of electric gadgets could be built in the bright new electrical world.

The only problem was that all these new gadgets needed power. Static electricity is good for electrocuting kite fliers and allowing kids to shuffle their feet on carpeting and then shock their siblings, but it is only when electric charge flows along conductors, such as copper wire, that it is useful.

What was needed was a way of getting all those electrons moving—to create electric current.

## Ohm time

The laws governing current, voltage, and electrical resistance were discovered in 1820 by a German school teacher and physicist named Georg Ohm, after the Italian Count Alessandro Volta had helped by inventing basic batteries.

Ohm found that a battery forces charge to flow through a circuit because it applies a *voltage*. The voltage of a battery is the difference in *electric potential* between its electrodes. Electric potential is just a fancy way of saying "amount of electric charge." So if you took your lemon battery and replaced your electrodes with metals that produced and absorbed electrons even better than zinc and copper (when you stuck them in a lemon), then your anode would be even more negatively charged and your cathode even more positively charged, so there would be a bigger difference in electric potential, and so the battery would have a higher voltage. Whew! Got that?*

The electron flow around a circuit is called electric *current*—more voltage, all else being equal, means more current. But the electrons don't really want to flow around the circuit—they were quite happy where they were—so a circuit is said to have *resistance*. The more resistance, the less current is allowed to flow.

You can think of the whole thing as being a bit like a bike wheel, with the rim and tire standing in for the electrons. The voltage applied by the battery is like the force applied by the pedals: It causes the whole wheel to turn (so all the electrons in the copper wires begin to move at once, at a speed of a few inches per hour). The resistance

---

*Another way to raise the voltage might be to use a stronger acid than lemon juice.

## Ohm's law

Ohm's law is expressed in this equation:

*Current (measured in amps) = Voltage (volts)/Resistance (ohms).*

This equation simply states that you can find the current flowing through a circuit by taking the applied voltage and dividing by the circuit's resistance. More voltage means more current; more resistance means less current.

of the circuit is like your hand slowing the wheel as it spins, with the rotational energy of the wheel being converted into heat by friction so that it warms your hand.

Ohm's discoveries are pretty straightforward. Ohm's law says that if you double the voltage across something that resists current flow (such as the filament of a lightbulb) you double the current (which is basically the flow of electrons through the circuit). It holds true as long as your circuit doesn't heat up. Then again, if you put current through a lightbulb, it does heat up, so double the voltage doesn't quite produce double the current.

Also important is the amount of power that the battery delivers to the bulb. This is given by multiplying the current by the voltage. So, if you can double the voltage and the current, you can quadruple the power delivered to the lightbulb. (Until your bulb burns out, anyway.)

A flow of charge in one direction, as in these examples, is called direct current (or DC). But, as physicists studied electricity and magnetism in the 19th century, one odd

---

### Electric Power

The equation for electric power is very simple:

*Power (measured in watts) = Voltage (volts) x Current (amps).*

As with other simple equations, this can be easily rearranged. So Current = Power/Voltage and Voltage = Power/Current.

---

genius, Nikola Tesla, who came from Serbia but settled in New York, would discover that it was much easier to generate and distribute another form of electricity, where the electrons don't flow around a circuit but instead rock back and forth in once place, creating alternating current (AC). But first, someone had to discover the link between electricity and magnetism.


## The attractions of electricity

In 1820 a Danish physicist, Hans Christian Ørsted, noticed that the needle of his compass moved when the electric current in a wire near it was turned on and off.

This was a very important discovery. It showed that, just like magnets, electric currents create magnetic fields. So, for example, the conductors that carry them can be attracted to some metals, such as iron; they can be repelled by other magnets or conductors that have the same polarity (just as the north poles of two permanent magnets repel each other); and they can make pretty patterns with iron filings on a sheet of paper as they move to follow the lines of force in a magnetic field.

The effect is even better if you wrap a coil of current-carrying copper wire around a short iron bar: When the current is on, you have a strong magnet, but when it's off, none at all. Electromagnets, as these devices are called, are very useful. For example, they make the fabric cones in loudspeakers vibrate (so you can listen to your favorite band and damage your hearing all at the same time), and they pull items containing iron out of household waste so it can be recycled.

The strange thing is that the movement of electric charge actually explains all magnetism. Permanent magnets, like the one in your compass, are magnetic because of the way the electrons in the needle are permanently moving in sync, a bit like they would be in an electric current.

## Signs of movement

The fact that electromagnets can vibrate speaker cones and pick up steel cans from garbage gives us a clue that electricity can be used to create movement.

In 1821 a great British experimenter, Michael Faraday, was the first to make use of the fact that the magnetic field around a conductor will cause another magnet to turn. But his demonstration of the effect required that the wire carrying the current be dipped in a bath of toxic mercury, which made it an unlikely candidate for the development of electric motors.

Faraday also discovered that the process worked in reverse. If you moved a conductor through a magnetic field, a current would start to flow in the conductor. What's interesting is that the strongest magnet in the world can't make a current flow unless you move it. But the faster you move that conductor, the greater is the current *induced* in the conductor. So spinning a conductor—a coil of wire is best—in a magnetic field is a great way to generate electricity.

All this theory helped lots of inventors improve on Faraday's ideas, until in 1869 Zénobe Gramme, a Belgian electrical engineer who struggled with reading and writing and couldn't do advanced math, invented the first generator that could make enough power to sell. And when Gramme accidentally connected one of his efficient generators (that was generating electricity) to another (that wasn't) and spotted that the shaft on the second turned, he realized that he'd also invented the first usefully powerful electrical motor.

Soon, the first commercial power stations were being built. But still, the electric age was not quite born. One man, Thomas Edison, was holding everything up. Part of the reason was that he, too, wasn't great at math. And part of the reason was that he owned a lot of patents and made a lot of money off of direct current. He didn't want AC rocking his comfortable boat.

Edison tried a lot of stunts, including electrocuting an elephant, to convince people that alternating current was dangerous. He also exploited and ignored Nikola Tesla, who for a while ended up digging ditches for Edison's company. But Tesla was a genius. He had done the math (AC is much more complicated than DC), and his AC designs—particularly of motors, generators, and power transmission systems—won out all over the world, even if no one wanted to build the death rays and another strange machines he proposed in his later years. Edison, who had plenty of other patents and had started movie piracy as a sideline, merely ended up rich (and famous as the inventor of the lightbulb—which he didn't invent).

So what was the big advantage of alternating current anyway?

## AC/DC: but only AC really rocks

Alternating current (AC) is much more complicated than direct current (DC). The voltage is constantly moving between positive and negative; the current therefore flows one way, and then the other. The AC that's piped into some houses causes the electrons in its wires to rock back and forth 50 times a second. This makes the math complicated and causes strange effects. For example in a thick conductor, AC flows only through the surface of the wire—so the kind of high-current conductors used by power stations are usually hollow.

But AC has one huge advantage over DC. Because it is constantly changing, it is easy to change the voltage with a transformer, and as we'll see, that makes it easier to transmit over long distances, as in, say, a power grid.

Once your power grid includes transformers, you can make use of the fact that high-voltage AC can efficiently be sent many miles through a national power grid at 400,000 volts before being stepped down by a transformer to the 220 volts that come into most U.S. homes. Why use such a high voltage in the power grid? So that the electric current in the grid is low.

Remember that electric power is the voltage multiplied by the current. So for a given amount of power, you can use a low voltage and a lot of current, or a high voltage

## Transformers

A transformer is a very simple device, consisting essentially of two coils of wire next to each other. AC current in the first coil (called the primary coil) creates a magnetic field, which induces an AC current in the second (or secondary) coil.

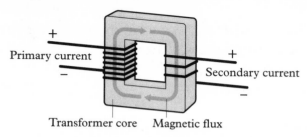

Primary current

Secondary current

Transformer core     Magnetic flux

You can't use a transformer to transform DC current, as it's the constant changing of the AC current that produces the magnetic field. All you will do is melt the fine wires in the coils, so don't try it!

and not much current, or something in between. But why is it good for the current to be low?

Let's go back to our bike-wheel thought experiment. Put the bike in low gear and turn the pedals—this is like using a high voltage to produce a low current. Now place your hand on the tire to act as a resistance. No problem, right? Now put the bike in high gear—to produce lots of "current" with a small "voltage"—and try again. Ow! Friction burn!

## How power stations work

In their most basic form, all power stations use a turbine to convert *kinetic energy* (energy that is due to movement—for example of steam or wind or water) into electricity. The source of the power—whether it's fossil fuels burned to make the steam, the wind, or water flowing through a hydroelectric dam—causes the turbine to move a large magnet around a coil of wire, which, according to the laws of induction, causes an electric current to flow in the wire.

It's the same with electric cables. All cables have some resistance, so put a lot of current through them and they'll heat up, throwing a lot of power away. Crank up the voltage, though, and you don't need so much current, so you reduce power losses from heating cables. And the power grid transmits a lot of power, so obviously it will use very high voltages to keep those currents down.

DC power stations, on the other hand, could be located no more than a couple of miles or so from the homes that used their power. Any farther away and they would lose too much power through cable heat loss. Either that or use high voltages that would be very dangerous in the home—at power grid voltages, electricity will happily jump short distances through the air.

So Tesla's AC systems won out, revolutionizing life. Starting in the early 20th century, you no longer needed one steam engine for every pump and train and machine. The electric age had begun.

# 8

# Relativity

*This chapter is all about special and general relativity, the concepts that Albert Einstein used to revolutionize our understanding of everything. Since Newton's time, people had thought the universe was neat and orderly, that it ran like clockwork. But Einstein showed that that's just how it seems. If you look carefully, everything is weird—which is why nuclear bombs go bang, and you can never beat the speed of light.*

A lbert Einstein was born in 1879. As a youngster he clashed with the authorities at school. At age 17, he changed nationalities to avoid military service. In 1900, he finally got a degree in math and physics. But he couldn't get a university job, so he became a patent clerk.

That hardly sounds like the résumé of a genius, but Einstein was stubborn. He knew he had some good ideas, and he didn't give up on them. His patent work helped because he looked at new kinds of electromagnetic devices, but most important, Einstein was very creative.

He knew that he had to think in a different way to solve the problems facing physics. And though the work of other physicists had given him many clues to follow, he couldn't rely on falling apples like Newton had. Physics had to go somewhere new. And so Einstein (like us) did physics by thought experiment, looking for general, logical truths about the universe, and trying to understand what they implied.

Then came 1905, Einstein's miracle year. After a day at the office, Einstein would go home to his wife, Mileva, a trained engineer, and to his 1-year-old baby son (though he admitted he wasn't a great family man). And there, within 12 months, he found the energy to write and publish four startling, revolutionary papers on physics.

And the first and most important principle that those papers presented was *special relativity*.

## You don't know where you're going

Relativity wasn't entirely Einstein's idea. For example, all of physics is based on the idea that the laws of physics are the same at all times and everywhere. And Galileo broadened this idea so that it covered steady motion.

In the 1600s, the church told Galileo that the Earth couldn't be rotating to the East at hundreds of miles an hour because (a) God put man at the center of the universe, and (b) if it was rotating at such a ridiculous speed, any falling apple would move down and to the West. In reply, Galileo told the churchmen to go away and shut themselves in a windowless ship's cabin—and there do a long list of experiments.

As long as the ship moved at constant speed, in any direction, Galileo argued, apples would still fall straight down. In fact, he said, once in the cabin, it is impossible to tell which way the ship is sailing, just as we can't know which way the Earth is rotating, other than by looking at the stars.

Others developed this idea that events are relative to the observers that see them. For example, just because you're in the Back of Beyond doesn't mean you're a long way away from me. I might myself be taking a short break just around the corner, in Beyond.

## It's a frame-up

Similarly, if you're driving down the highway and pass a police car at a speed of 22 m.p.h. (10 m/s), you might wonder why it comes after you, sirens blaring, and you get a hefty fine and maybe three points on your driver's license. But of course if the police car was already doing 68 m.p.h. (30 m/s) when you passed it, that means, as their clever on-board camera shows, that you were actually doing 90 m.p.h. (40 m/s).

In physics at least this is a relatively uncontroversial bit of relativity. But that doesn't mean that all of relativity has always been uncontroversial. Back in 1905, Einstein was about to tear apart Newtonian physics and make relativity much more interesting.

## Lighting the way

As the 1800s ended, it was becoming clear that light was special. It wasn't just that no one could see the ether through which light was meant to travel, and that light's speed (in a vacuum) was always the same: it was also becoming clear that the speed of light itself was important, for example because various equations suggested that it is impossible to travel faster than light. So scientists started looking more closely for evidence that proved ether existed, and to see how it might affect the speed of light.

And so Einstein took a bold step. He ditched the idea of ether and, building on the work of others, made the constant speed of light the centerpiece of special relativity.

First of all, it's worth realizing just how weird this idea is and how it messes up the nice solid Newtonian universe we know and, on pleasant days in the apple orchard, generally love.

Let's get you off Galileo's boat and onto a train, because Einstein used trains in his thought experiments. Let's say your train is moving east at 75 m.p.h. But that doesn't matter to you. Whether you throw your apple forward or back, it moves away from you at 25 m.p.h.

Now, as good old Albert watches from the platform, he sees your apple move at different speeds backward and forward. Relative to him, when you throw your apple forward, it moves forward at 75 m.p.h. plus 25 m.p.h., which is a whopping 100 m.p.h. But when you throw it backwards, it merely moves forward at 25 m.p.h. less than the train's speed; that is, at 50 m.p.h.

All this is fairly obvious. But Einstein said that light is different. If our apple, instead of acting like an apple, acts like a photon of light, it doesn't matter which way you

## What's the ether like?

Everyone knows that sound waves travel through air (and other substances) and water waves travel through water—but what do light waves travel through? What is this invisible stuff—which physicists used to call *ether*—that brings the sun's light and energy to us? Toward the end of the 19th century, Albert Michelson and Edward Morley spent six years carefully building an intricate device to investigate the ether and answer this question.

They knew that the Earth is speeding through space, spinning and circling the sun. And so it should be travelling through the ether, too. In one direction, therefore, the speed of light should seem faster relative to a scientist on Earth, just as the wind seems to blow harder if you run against it. And in the opposite direction the speed of light should be slower.

But when Michelson and Morley ran their intricate experiment, it failed completely. They tried again and again, over a period of years, but with no luck. Whichever direction they chose to look in, the speed of light was exactly the same. There was no evidence at all for the existence of ether.

throw it: it always passes Albert, who's still there on the platform waiting for his train, at 75 m.p.h. What's more, relative to you, every photon of light you throw out goes away from you at 75 m.p.h.

If you don't understand this, don't worry. It makes no sense with apples. Let's try it with those little packages of light known as photons.

## Time to wait

Poor old Albert's been waiting around on that platform for a long time. And as we'll soon see, he has discovered that time isn't quite as dependable as most of us think. If, for example, that train platform is somewhere in outer space near a black hole, he could be there approximately forever without seeing a train.

So let's give him something useful to do: another thought experiment. We need two tall glass boxes—one for you and one for Albert—each with two mirrors inside, one on the top face and one on the bottom. And in those boxes, bouncing in sync between these mirrors, you each have your own tiny photon of light.

Now, if your train is stopped at the platform, and because you and Albert have fantastic eyesight, you can see your photons bouncing up and down together. So far, so simple. But what happens if we try the same thing while your train is moving at 90 percent of the speed of light?

## On the fast train

Here you come. (Don't get your hopes up, Albert. It's not your train.)

As you look at your box, your photon is bouncing up and down as before. No problem. But what does Albert see? He sees your photon move diagonally, in a series of zigzags. It has to, because after it leaves the middle of the bottom mirror, it travels up and hits the middle of the top mirror—and that mirror (along with the train and you and the box) has moved forward a bit as it traveled with the train.

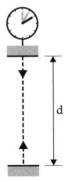

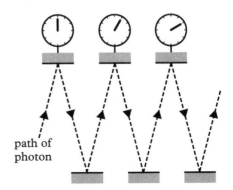

The motion of Albert's photon as he stands on the platform

path of photon

What Albert sees your photon do as you move

But this is where it gets interesting. That diagonal path that Albert sees your photon take is like the long side (hypotenuse) of a right triangle, because it has gone up and across a little bit. And, of course, this long side is longer than the path you see the photon take because for you, it only goes up. In fact, if your train is moving at 90 percent the speed of light, that path is roughly twice as long.

This is much the same as what happened when you were throwing your apples forward at 25 m.p.h. just now. If something moves a greater distance in the same amount of time, it must be going more quickly. So if your photon was an apple, Albert would simply see it move faster than you see it move.

Except, of course, that light can't go faster. And so something has to give. As Einstein discovered, that something that gives is time (and actually distance, too, but we'll just worry about time for now and ignore the fact that superfast travel makes you shrink a bit). Time gives up being constant. Because you're moving relative to Albert and because, as Albert sees it, your light has farther to travel between bounces (and because it has to travel at the same speed), *time for you must run more slowly than for him.*

Of course, you can't tell that time is running more slowly for you. Your watch still seems to tick as constantly as ever. But if your train is going past Albert at 90 percent of the speed of light, as he sees it, for every *five* bounces of the photon in your box, the one in his bounces *ten* times.

And it's not just an effect on photons. Deep in the laws of physics (and Einstein did do the math), time really does slow down as you speed up. So, according to Albert's measurements, your watch would tick more slowly, and if he could see you doing a little dance of physics happiness in time with your bouncing photon, he'd see you doing it all in slow motion.

Even at low speeds this happens. It's just that the effect is so small that we can't see it. But if your train is moving at 75 m.p.h. and you throw your apple forward at 25 m.p.h., and Albert on the platform measures its speed with his new Super Accurate Apple Speed Measuring Device (which he's just invented), he'll see it moving forward at a tiny bit less than 100 m.p.h.

So once you accept special relativity, you have to accept that many other things in physics change.

## Faster than the speed of light?

Your fast travel slows down your time, relative to stationary Albert on the platform—but relative to you, on the train, it's Albert's time that has slowed. One result of this confusion is that some things, which look like they're

## That famous equation

The equation that tells us how energy and mass are related—and how much energy is hidden in all the solid stuff around us—is the most famous equation in physics, which Einstein discovered in 1905:

*energy = mass x speed of light squared, or as it is more commonly written:*

$E=mc^2$.

happening at the same time to you, will look like they're happening at different times to Albert.

Hold on a minute, you say. This makes no sense and it's all based on not being able to go faster than the speed of light. But who says we can't travel faster than the speed of light? In your thought experiment you've built a super-powerful rocket. And here you are, blasting off into space. What stops you from going faster and faster and faster—to the speed of light and beyond?

Well, one way to look at what happens is to look at kinetic energy: the energy a thing has due to its movement. Kinetic energy is related to the mass of an object and its speed squared (see pages 42–44).

According to relativity, the faster you go, the more difficult it becomes to go even faster. But you're still adding kinetic energy with all that rocket power you're using. So if your kinetic energy is increasing and your speed isn't—then your mass must be.

At low speeds, which in relativity means speeds less than a third of the speed of light (which, you'll remember, is nearly 670,000,000 m.p.h., or 300,000,000 m/s), it's almost impossible to notice this effect. But as your rocket's speed increases beyond 224,000,000 m.p.h., or 100,000,000 m/s (much, much faster than any human being has ever gone), all your extra pushing just makes your rocket heavier and heavier and even more difficult to push. And that's why you and Albert and I can never travel faster than the speed of light—even in a thought experiment.

## New killer power

Once you've discovered that $E=mc^2$, it is obvious that all you need is a way to convert mass directly into energy, and you will get a very large amount of energy from a very small amount of mass (because the speed of light is very large). Then we can stop drilling for oil and digging for coal and warming the planet, and make as much energy as we need, because we can always spare a bit of mass—especially around the waistline.

Unfortunately, controlling that mass-into-energy conversion is tricky, especially since once you find some way of getting the process going, lots of that energy tends to come out as highly dangerous forms of radiation.

So it's easier to make nuclear bombs (which convert some mass from the nucleus of an atom into a huge, energetic bang and a lot of dangerous radiation) than it is to make a nuclear power station, which also makes energy from mass, but also, unfortunately, makes radioactivity.

The bomb dropped on Nagasaki at the end of World War II converted just 1 gram of mass into energy. But it killed 40,000 people immediately, and another 40,000 died from injuries and radiation in the following few months. And many more died of cancer and other diseases caused by radiation in the years since 1945.

## Floating and falling are the same

Making relativity special (and Einstein's other 1905 stuff that we'll find out about in the next chapter) might have been enough for some scientists. But Einstein wasn't satisfied. By 1907 he was making progress on a new, more general kind of relativity theory that included gravity. He called it *general relativity*.

Newton had done a good job with gravity. He used the movement of the planets to come up with mathematical descriptions of gravity that are still useful today. And he realized that the force of gravity was a universal. Just as it made the apple fall toward the Earth, it made the moon fall toward the Earth (except that the moon keeps on missing because it's going pretty quickly, and so it ends up flying around and around). In the same way, the Earth and the other planets in the solar system fall toward the sun, but keep missing.

But what Einstein realized in his thought experiments (and published in his 1916 paper when he had it all figured out) was that there is no difference between what the moon or a skydiver feels in free-fall—and what you'd feel if we stuck you in a spacesuit and sent you out of our galaxy into empty space, where gravity is very weak. (Except that when you're out alone in the space beyond our galaxy, you may feel a bit worried about how you're going to get back to Earth, given that you can't travel faster than light and, even at the speed of light, it would take you 30,000 years to get home.)

The similarity of free-fall and floating where there is no gravity becomes clear if we ignore the effect of wind resistance and consider what happens to apples dropped by the skydiver and by you in deep space. What happens is that those apples just seem to sit where you let go of them. Alternatively, if you or the skydiver gives them a push, they continue on in whichever direction they were pushed.

So, what Einstein realized is that when you're in free-fall and accelerating along with gravity, it is impossible to see any effects from gravity. It no longer affects you at all.

And then he came up with another thought experiment.

## Sitting down is like blasting off

What, Einstein wondered, is the difference between sitting in a comfy chair in the windowless cabin of Galileo's ship (which is moored in the harbor) and sitting in a similar comfy chair in your similarly windowless spaceship that is accelerating at 9.8 m/s² as you begin that 30,000 light-year journey back home from the Back of Beyond? (You've chosen to accelerate your spaceship at 9.8 m/s² because it's the same rate that the Earth's gravity accelerates falling apples and everything else.)

And the difference that Einstein came up with is that *there is no difference* that you can detect. For example, in both places, if you get out of that comfy chair and onto a set of bathroom scales, they will say that you weigh the same. And an apple you drop in either place will pick up speed at exactly the same rate: $9.8$ m/s$^2$.

Einstein explained this with lots of complex math, but you can get a good idea of his results when you understand that mass, instead of making an invisible force called gravity, instead makes space and time curve. Every object in the world curves space and time, and big things curve it more, and these curves are like pits around the objects that create them. The curvature of time and space—the pits, as we've labeled them—then cause other objects to move (as if they were rocket-powered) because they slide into them.

## Falling into bed

One way to picture the curvature of time and space and the movement that it causes is to think of space as a cheap old mattress on a cheap old bed, the kind that sags a lot when you lie on it.

Whoever gets into bed with you will enter your gravitational field (which is the bit of space and time—that is, mattress—that you've curved). Once that person is on that curved part, they'll slide into your pit and there'll be two of you down there, and with their mass added that gravitational pit will be even deeper and more difficult to struggle out of than it was.

If a couple of children, the dog, and Grandma then also climb on the bed and roll into the pit you've made, then together you've pretty much created your own black hole: an area of space and time so curved by a huge and compact mass that it is impossible for anything—even light—to escape its gravity.

## I don't believe it!

And neither did many scientists in the early 1900s. After Einstein, nothing was quite what it seemed anymore.

In Einstein's math, instead of space and time being, well... space and time, they are just aspects of a single curved, four-dimensional system, which includes our three usual dimensions (up/down, left/right and in/out) along with time. The system sucks up the force of gravity, too. So, instead of the moon orbiting the Earth because of the pull of gravity, the moon (as far as the moon is concerned) has no forces acting on it and is continuing along a nice straight line. It's just that with space and time curved by gravity, that nice straight line is basically a squished circle.

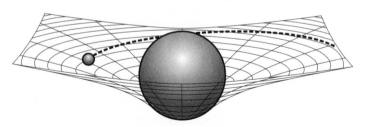

How space is curved by a massive object

And mass, for such weighty stuff, is no longer solid, dependable, or necessarily that weighty. Mass, energy, and—it turns out—momentum, are simply three different ways of looking at the same thing. So all three are sources of gravity. (And, to make it even more complicated, the related quantities of internal pressure and tension are also gravity sources.)

We've already seen how going fast makes you massive. So does being energetic or under pressure. It's almost as if you go on a diet and turn all your excess weight into energy and higher blood pressure and find that, because energy and pressure have mass, you weigh exactly the same as before you took out that expensive gym membership.

Yes it's complicated and unintuitive, and some people didn't like it, but the predictions that Einstein's theory made, and the way it agrees with all kinds of measurements and experiments, are remarkable. It simply has to be right.


## Bending starlight

Newton's law of gravity predicts that the planets orbit the sun in perfect, fixed ellipses. But general relativity predicts that those ellipses themselves will slowly rotate around the sun. This effect had been noticed in the orbit of Mercury before Einstein explained it, and soon after Einstein published his new theory, astronomers were able to start confirming that all planets in the solar system had relativistic orbits.

That was only the start of the evidence. Einstein predicted that intense gravitational fields bend light—and in 1919, when this deflection was measured, Einstein became an international star and part of popular culture. Meanwhile, a number of competing theories from other eminent physicists were discarded for good.

General relativity also predicted that gravity—like movement in special relativity—affects the passage of time. The more intense the gravitational field around you, the more slowly time will run for you. Again, this isn't saying just that your clock will run more slowly—though it will—it is saying that time for you really is slower than for someone feeling less gravity. Time spent in a high gravity

## General Relativity and GPS

The strange result of gravity on time is well proven. For example, compared to the interminable time you experience while stuck in a traffic jam, time literally runs faster (because gravity is weaker) in the orbiting GPS satellites that help your GPS system get its fix. GPS relies on very accurate timing, so the clocks in GPS satellites are built to take the physics into account. On Earth, they tick 39 microseconds a day more slowly than accurate Earth clocks. Then, when they're in orbit, they tick at exactly the same rate as Earth clocks.

field is a bit like time spent in a bar doing physics on a pool table. You think you've been in there barely half an hour, with hardly enough time to finish your experiment, let alone your drink, but outside it's 6 hours later, it's cold, it's dark, and your dinner has been fed to the dog.

## Black holes and a big bang

Besides making the specific predictions that we've looked at, general relativity turned out to apply really well to the universe as a whole. Once it was confirmed, it revolutionized cosmology and astrophysics, the sciences that try to explain the entire universe and how it came to be here.

For example, general relativity is an integral part of the theory that says the universe was born in the Big Bang. It also predicted black holes long before we'd built the technology to go looking for them.

A black hole is a place—a collapsed star or a bunch of them, squeezed together—where intense gravity slows time so much that if you were to fall into one, we'd see you move more and more slowly as you approached what's called its *event horizon*. But even if we watched forever, we'd never see you actually cross it and fall into the hole. Instead, the light from you would slowly fade and change (it would, in fact, be *red-shifted* as if you were moving away very quickly; see page 157).

However, that doesn't mean that a quick dive into a black hole offers you a way to take time out and come back in a few thousand years. Remember, relativity says that you can't see time speed up or slow down for yourself. So you'd cross the event horizon without noticing anything odd, except that you wouldn't be able to turn around and escape. And then...well, we're not sure exactly what happens then. But with all that gravity, it's likely to be a crushing experience.

## Newton versus Einstein

All in all, Einstein's theories of relativity showed how the laws that Newton and many other classical physicists had discovered were merely *just about* right. And in physics, just about right is, in fact, wrong.

Nevertheless, you can't argue with the fact that Newton created modern physics. Newton's system described the behavior of the entire cosmos, and while others before him had invented grand schemes, Newton's was different. His theories had math. They made specific predictions that could be confirmed by real experiments. In a way, Newton was lucky: he made such a big impression because he was first. And in a way, it doesn't matter that we now have better versions of his just-about-right laws. Engineers and some other scientists are quite happy with just-about-right. Newton's laws, and lots of the math that he figured out, are still used regularly.

Unless you start traveling faster than about a third of the speed of light, or shrink down to the size of an atom, Newton's laws cover just about all the running, jumping, throwing, and hitting you can get away with. Where they fall down is if you want to understand what makes the universe tick and where it comes from.

In this regard, after 1905, everything in physics was different (though it took some physicists many years to accept the new ideas, and some never did). Newton's reign as chief physicist genius of all time was over. He couldn't compete with Einstein, a man who almost single-handedly changed how we think of space and time. Especially when, as we'll find out next, Einstein went on to almost single-handedly change how we think about matter and atoms, too.

# 9

# Quantum Physics

*In the first few decades of the 20th century, physicists made huge leaps forward in their understanding of light and atoms. Einstein started everything off, but this time there was also a host of other brilliant men who helped us realize how, if you look closely, the world is unbelievably odd: Elements have fingerprints, energy comes in indivisible chunks, light is both a wave and a particle, and some things can be in two places at once.*

At the start of the 20th century, as we've seen, physics was in a mess. The speed of light was one big problem. But there was also a little problem that was just as big: the atom.

Chemists had produced good evidence for the existence of atoms. And at everyday temperatures, as we've seen in chapter 5, Newton's laws could be applied to large numbers of atoms and molecules to predict the behavior of gases and solids.

But throughout the 19th century, physicists had been discovering areas where the combination of Newton's laws and the theory of matter (that said it was made up of little bits called atoms and molecules) did not work. For example it would predict that if your toilet seat was really cold (around absolute zero), its specific heat capacity would be so low that it would warm up instantly as you sat on it. And we all know that can't be true.

These problems led many people to believe that atoms were not real. Chief skeptic was an Austrian physicist and philosopher, Ernst Mach. He argued that, because atoms couldn't be seen, they didn't exist. For him they were no more than an occasionally useful idea.

But Mach was wrong. Clever arguments and fancy philosophy are no match for physics and experiments. And they certainly weren't a match for Albert Einstein.

Yet in what came to be called quantum physics, Einstein outdid even himself. Some of the ideas about atoms and light and matter that he and a few other very clever physicists discovered are so strange that even Einstein could never really get his head around them. He thought his theory went so far against common sense that it was just a stepping stone to something more sensible, and he spent many unsuccessful years trying to find it.

So don't worry too much if what we discover in the next few pages seems a little odd.

## You can't be serious

Einstein was a realist. He believed that there must be a single theory that could prove the existence of atoms and accurately predict how groups of them would behave.

And, as the greatest genius in physics, he had the brain-power to start creating it.

All of Einstein's 1905 papers in the German scientific journal *Annalen der Physik* caused debate and argument. In the second paper, Einstein did the math that made sense of Brownian motion. In the third, as we saw in the last chapter, he announced his theory of relativity. In the fourth, he discovered that mass and energy were related by the famous equation $E=mc^2$ and paved the way for nuclear bombs.

But it was Einstein's first paper of 1905 that most shook up physics and physicists all over the world. It was not-very-snappily called "On a Heuristic Viewpoint Concerning the Production and Transformation of Light" (*heuristic* means explorative), a title that highlighted the fact that even Einstein wasn't entirely convinced by his reasoning. And in it, he set forth the revolutionary proposal that light can be two different, contradictory things at once.

## Don't open that oven—it's a can of worms

In 1900, a couple of British physicists realized that the accepted laws of physics predicted that if you looked into a hot oven, you'd immediately be burned to pieces. This was obviously wrong, and Einstein's first paper of 1905 started to explain why.

The only problem was that this explanation seemed more impossible than the paradox it was meant to solve. It took nearly 20 years before the majority of physicists began to accept it.

## Black bodies

A black body is a physics term for an object that completely absorbs all frequencies of electromagnetic radiation. Black bodies exist only in thought experiments (although we can build close approximations), but thinking about them produces the math that helps us figure out things like the temperatures of stars or how the greenhouse effect is affecting the climate.

As a black body gets hotter, it emits higher frequencies of electromagnetic radiation. At first that radiation is invisible, but from a few hundred degrees C up, it appears red, then yellow, and so on through the visible spectrum.

You can see the same effect with a regular lightbulb on a dimmer, because the bulb's filament acts like a black body. At minimum power, the light from the bulb will appear somewhat red. Slowly turn the power up, and as the bulb gets hotter, the light will turn orange, then yellow, then eventually white at full power.

The biggest problem was that Einstein's new theory suggested that light (and all other electromagnetic radiation) is both a wave and a particle. For a realist in particular this was a terrible result, because waves do wavy things (like pass through each other unharmed), and particles do martial arts on pool tables.

It was about as bad as telling physicists, "I've found a theory that says something can be in two places at once." In fact it turned out that in a way, this was exactly what Einstein was telling them. No wonder so many of them hated the idea.

## Chunks of energy

It makes sense that a hotter object makes a brighter light because, as we know, hotter things have more energy. So a 100-watt lightbulb is brighter than a 40-watt. But why doesn't the radiation come out of a black body in equal amounts at all frequencies, so that to our eyes, it always looks white, whether it's dim or not? Why is it that at lower temperatures, lower-frequency radiation is more common, so the black body looks red?

The man who came up with the answer—even though it made no sense to him—was German physicist Max Planck. He spotted that the only way to make the math work was to assume that the vibrational thermal energy of atoms can be released only as light energy in chunks of certain sizes, and that at higher frequencies these chunks would have more energy, because higher-frequency electromagnetic radiation carries more energy around.

The upshot of Planck's discovery is that if you have a quantity of light energy and keep dividing it in half,

---

### Planck's constant

Planck also discovered that the size of these tiny chunks is given by multiplying the frequency of the light by a very small number, which came to be called Planck's constant. (Physicists write this as $E = hf$.) Planck's constant is usually represented by the symbol $h$, and its value is 0.00000000000000000000000000000000 06626 Joule seconds (usually written as $6.626 \times 10^{-34}$ Joule seconds).

eventually you reach a small amount of energy that cannot be divided any further. Planck called this tiny chunk of energy a *quantum*.

Now, at high frequencies these chunks become quite large, because the frequency of the radiation becomes large. This is even clear in visible light: the deepest violet light we can see has double the frequency of the darkest red; so chunks of violet light must have double the energy of chunks of red light. Once you know this, it is clear that unless our black body (or bulb filament) is very hot, very few of the hot little atoms in it will have enough thermal energy to make chunks of blue light, and so red light is more common. (Remember that all the atoms in a hot black body have different amounts of energy, but it's the average energy of all the atoms that gives the body its temperature.)

So by inventing the idea of the quantum, Planck came up with a partial explanation for why hot bodies glow red and really hot ones glow white. But it was only after Einstein got ahold of the idea that it became part of our picture of how the world works.

## Photon torpedoes

If you shine light and ultraviolet on to a piece of metal, the light can knock electrons out of the metal. In a physics lab, these can easily be collected and the number being knocked out can be measured.

Back in the 1900s, this photoelectric effect was making physicists rather uncomfortable. The wave theory of light suggested that as you turned up the brightness of the light, you should eject more electrons, because brighter light has more energy.

But in their experiments this didn't happen. If you shine red light onto a piece of potassium, no electrons are ejected, no matter how bright you make the light. But if you change the light to violet, you can turn it down low and still electrons will be ejected.

Einstein had the genius to see that, if light was made up of tiny quantum particles (which we now call photons), the photoelectric effect is easy to explain. As Planck had shown, it's the frequency, not the intensity, of the light that determines its energy. So Einstein's 1905 paper suggested that low-energy red photons don't have enough energy to knock any electrons out of the potassium, no matter how many of them you throw at it. But even a single ultraviolet photon has enough energy to knock an electron out.

## Testing times

This explanation of the photoelectric effect may seem obvious now, but at first, hardly anyone believed Einstein. One American physicist, Robert Millikan, was so annoyed and so wedded to all the evidence that light must be a wave that he spent *10 years* carefully testing the photoelectric effect in order to find something wrong with Einstein's theory. But his careful experiments only ended up confirming Einstein's ideas in every single detail.

For this useful work, Millikan won the Nobel Prize for physics. Which just shows that in physics, you can get it right even when you're wrong.

Slowly but surely, physicists began to accept Einstein's radical ideas. It helped that other brilliant thinkers used and developed the idea that light comes in packets called photons to explain many familiar effects, including how light can be produced by various substances even when they're cool, or at least not very hot. (A few examples are neon lights, glow sticks, and the Northern Lights—where the dark skies above the magnetic north pole glow with spectacular green and red curtains of light.)

## Fingerprint physics

All elements, when turned into a gas and heated, give out their own distinct color. And if, using a prism, this color is separated out into a spectrum, there will be bright colors in some places and no color at all in others: a fingerprint that is unique to every element in the universe. This is particularly good news for astronomers, because it means that simply by looking at the light that the sun and other stars send us, they can learn a great deal about the substances that make them up.

The simplest case is hydrogen (because, it turns out, hydrogen is the simplest element—it has just one electron scooting around just one proton). If you heat hydrogen until it glows and then, using a prism, spread out the light it produces into what would be a rainbow (if it were from a white-hot black body), all you see are four bright lines of color: red, blue-green, blue, and violet. The rest of the rainbow is dark.

But if you then let your hydrogen cool, and shine a white light through it, and then through the prism, you get almost the whole rainbow of colors—except with thin dark lines at exactly the places where the bright bits were at the time the hydrogen was heated.

So hot hydrogen emits red, green, blue, and violet light, and cool hydrogen absorbs that light.

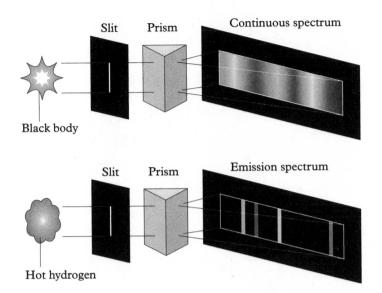

In 1885, when Swiss schoolteacher Johann Balmer looked at the frequencies of these four bands of light, he realized that they fit a pretty simple mathematical formula. (He was also able to predict that other frequencies of electromagnetic radiation would be emitted—as ultraviolet and infrared, and so on—and years later, physicists with new technology discovered these.)

Because Balmer's equation was so simple, it seemed likely that it had something really important to say about what happened inside atoms. The only problem was that nobody knew what that thing might be for another 27 years, until Danish physicist Nils Bohr arrived in Manchester, England, to work with Ernest Rutherford, the man whose experiments had given us a picture of the atom as a tiny positive center surrounded, at a large distance, by negative electrons.

## Inside the atom

After J. J. Thomson discovered the electron in 1897, it was thought that atoms were like a "plum pudding," in which electrons (the plums) were surrounded by a "pudding" of positive charge, which held them together.

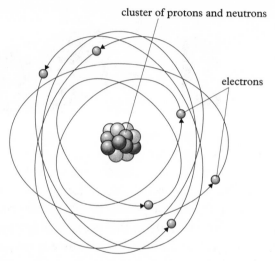

cluster of protons and neutrons

electrons

Carbon-12, with six protons and six neutrons in the nucleus

This changed in 1909 when Rutherford, a New Zealander who came to be known as the father of nuclear physics, directed alpha particles at a thin sheet

## Great explanations

But back to Bohr. The Dane had arrived in England in 1911 with little English, but a big dictionary, the complete works of Charles Dickens, and the ability to work very hard.

of gold foil and measured how they were deflected. Under the plum pudding model, most of the alpha particles should have been deflected just a little as they made their way through the pudding of gold atoms. What actually happened was that most went straight through with no deflection, but a very small number bounced almost straight back.

This shocking result led to our understanding of the atom as a small, dense, positively charged nucleus orbited by electrons. Rutherford did the math and found just how little space the nucleus takes in the atom—it's so small that, if you could crush up all the human race, and do away with all the electrons, and overcome the huge repelling force of all those positive nuclei, you could fit us all into a block the size of single sugar cube (though he didn't put it quite like that).

The nucleus itself is made up of smaller particles—protons (which are positively charged) and neutrons (which have no charge). The number of protons determine what element the atom is, while the number of neutrons affects how stable it is—too many or too few neutrons and the atom may break apart or *decay*, resulting in radioactivity and radiation.

It didn't take him long to revolutionize our understanding of the atom. He knew that electrons couldn't just keep floating around the atomic nucleus by themselves—there needed to be something to stop them from radiating away all their energy (as light,

ultraviolet, etc.) and falling into the middle with a uni-
verse-ending bang.

His first brainstorm was to combine Einstein and
Planck's idea of energy chunks with Balmer's equation
that described the emission spectrum of hydrogen. His
idea was that electrons can sit only at certain fixed quantum distances from the nucleus. When they get more
energy (from heat, for example) they become "excited"
and move further away from the nucleus, say from level
one up to level four. (It's a bit like your having more
potential energy when you're sitting upstairs.)

But like you, electrons can remain excited (or upstairs)
only for so long. After a while they fall back to a lower
energy level closer to the nucleus (falling, say, from level
four to level two), and releasing their extra energy as a
chunk of electromagnetic radiation (in this case as a photon of blue-green light).

Because the various energy levels that an electron can
sit in are fixed (like the stories in an apartment building are
fixed), the packets of energy emitted as they move between
them are fixed, too—and therefore so are the frequencies
and colors. In hydrogen, for example, the fall from level
four to level two always gives off that blue-green photon.
And this creates the colorful elemental fingerprints that
physicists and astronomers see in their experiments.

## Exciting times

Bohr's idea also explains why gases absorb certain frequencies of light. It's like emission, but in reverse. As the
light photons hit the atoms, only those that have the right
amount of energy (and therefore the right frequency of

light) can be absorbed. So if a hydrogen electron is on level two and a blue-green photon hits it, it has just the right amount of energy to be absorbed, and the electron will jump up to level four. Yellow photons have the wrong energy for hydrogen electrons, and pass straight on through. (It's a bit like a crowd at a basketball game. Only the blue-green team's baskets get the blue-green half of the crowd excited. They don't even bother with polite applause when the yellows do something good.)

Einstein and Bohr, on the other hand, did receive polite applause for their work on quantum theory. And so did the increasing band of geniuses around them. But this was faint praise for the enormous and complicated advances made in physics as quantum theory was rapidly discovered and refined between 1900 and 1928 (especially when you consider that many of the contributors—including Werner Heisenberg, Erwin Schrodinger, and Paul Dirac—are definitely in the top 10 physicists of all time).

For example, Lederhosen-wearing, mountain-climbing German genius Heisenberg discovered that once you start poking around at the quantum level, there are real boundaries to what you can find out about photons and electrons. His *uncertainty principle* says that if you find out where exactly a photon is, then you can't tell how fast it is going (and vice versa). And this is not because our measuring systems aren't good enough. Instead, measuring one quantity (speed, for example) must affect the other (position, for example), so it can't be known exactly.

Part of the reason that these men didn't receive the worldwide fame they deserve is that the more you learn about quantum physics, the weirder and more difficult to grasp it becomes. And partly the problem is that the experiments that support it aren't easily explained either. But in 1986, all that changed. Laboratory technology was finally good enough to carry out the most important experiment in quantum physics—the one that, almost on its own, makes sense of the whole nonsensical subject.

## Quantum physics laid bare

I'm sure you remember our day on the beach: the sun, the surfers, the double slit experiment, which seemed to prove that light is made up of waves, not particles, because the light waves interfere and cause a diffraction pattern. You even made diffraction fringes in the gap between your fingers.

But what happens if you try the same experiment with single photons of light? If you take a very dim laser beam that sends out those single light chunks, one at a time, and you let them pass through two slits, what happens then? At this, its barest, loneliest point, does light behave like a wave or like a particle?

Well, the clever, amazing answer, the one that brings the abstractions of advanced quantum physics into the real world, is that light really is a wave and a particle: both, and at the same time, too. And if you poke at it to try and find out more, well then it just collapses and is one or the other—depending on what you do to it.

OK. Here's the experiment. We set up our dim laser, our narrow slits, and behind them a piece of sensitive photographic film, so that the arrival of each individual photon of light will be recorded. We turn our apparatus on. The first photon is produced. It speeds through. Bang.

It hits the film and leaves a mark. A single mark. It is after all a single photon. Aah! It's a particle!

But where is that mark? Well, you can't predict where it will be. But it is unlikely to be in a straight line from the laser, through a slit, and onto the film. Somehow that confounded particle has interfered with itself and not travelled in a straight line. What's more, if you leave the apparatus running for a few hours, the pattern that appears on the film is the familiar fringes, those exact same regions of dark and light that prove that interference is taking place.

And whether you run the apparatus for hours and let a million photons through one at a time, or turn up the light and blast them all through in a microsecond, the pattern is the same. More photons arrive where the light interferes constructively, fewer where it is out of sync, and cancels itself out.

The answer is inescapable. Each lonely photon travels through both slits at the same time and then interferes with itself, just like any good wave! Photons can be in two places at once, BUT (and this is a very big and important but, as you can see) if you start checking for them at the slits, you find that they really do go through one slit or the other.

In a way, it makes no sense. But it is what happens. Maybe there is no more to say than that.

## But what does it mean?

In quantum physics, meaning is hard to come by because things seem so odd to us. What, we wonder, does it mean for everything to be both a wave that goes through both slits and a particle that goes through one? What does it mean when you say that we can never be sure about where a particle is and how fast it is moving?

It doesn't help that quantum physicists themselves aren't in complete agreement about how to explain their discoveries. The Austrian, Erwin Schrodinger, got so excited by some of his many love affairs that he came up with lots of important quantum math. But in an effort to challenge the way we see quantum physics, he also came up with his famous thought experiment, in which a poor cat, shut in a box and at the mercy of a chancy quantum event, is both alive and dead until a physicist opens the lid and forces it to be one or the other. For years, we were expected to believe that a cat could really be both alive and dead. Today, though, Schrodinger's cat is being retired. Many physicists are coming around to the commonsense idea that, although quantum events can defy logic (so our single photon can go both ways through the apparatus at the same time), cats can't. They are dead and leaving their hair in one place or else they are alive and leaving it everywhere.

Of course, our problem with quantum physics may be due to the fact that it is only 100 years old and most of us aren't familiar with it yet. Maybe it's like a new pair of shoes that haven't been broken in. After all, it took centuries for most people to accept that the world is round and that it revolves around the sun, despite there being easily available evidence for those theories even before we invented frequent flyer programs and space travel. (For example, when a ship comes over the horizon, it doesn't just appear as a dot and grow as it would on a flat planet. Instead, you see its masts first, and slowly the rest is revealed.)

## Really weird

Scientists certainly have few problems with the effects of quantum physics. They are still adding to the long list of things that it explains, in physics, chemistry, and even biology, where it is shining new light on how our brains work. Paul Dirac used it to predict the existence of anti-matter before it was discovered in the laboratory. And engineers use the advanced equations when they're building microchips and working on new kinds of computers. We even know that basic light and power switches wouldn't work without quantum physics (and an effect called quantum tunneling).

So quantum physics is useful and well proven. The only problem is, it still raises questions that even the best and most brilliant physicists can't always answer.

# 10

# The Universe

One of the great things about physics is that in spite of all the wonders we've discovered (and in this little book we have barely scratched the surface), there is still lots we don't know. There's also lots that we sort of may know but aren't quite sure about: stuff that causes debate and argument at physics meetings and on the Internet—and that leads physicists to spend piles of money building bigger and better particle accelerators so that they can make even bigger bangs and smaller particles when they smash things up.

What's more, most of these hotly debated questions are so hotly debated because they touch on important and fascinating theories about the history and origins of the universe and our place in it. And that's two excellent reasons these questions are worth a look.

So, let's finish up by thinking about, among other things, the causes of the Big Bang, the possibility of time travel, and the makeup of dark matter: the questions that can flummox even a top physicist.

## How big is the universe?

Our sun is one among hundreds of billions that make up our Milky Way galaxy, a vast cloud held together by gravity, with maybe 100 billion stars, plus more clouds of gas and dust, and, we assume, many hundreds of millions of planets (though, because planets are small and dark compared to stars, we've only been able to spot around 500 so far).

On a clear night, far away from the bright lights of the city, you can often see the Milky Way. It's an arch of white light across the night sky, with its center around the constellation Sagittarius (the one that's supposed to look like a mythical half-man-half-horse centaur, preparing to fire a bow and arrow). The milkiness of the Milky Way is made up of those hundreds of millions of far-away stars, and we see them form that milky band because the Earth is on the edge of things, near the edge of the flattened disc of our galaxy, in a side road off one of its four major spiral arms.

Our galaxy is huge: It is around 100,000 light years across. In comparison, the sun is just 8 light minutes away—and you could fit a million Earths in it. The next nearest star, Proxima Centauri, is 4 light years away. But, the next nearest galaxy, Andromeda, discovered in 1925 by the U.S. lawyer-turned-astronomer Edwin Hubble, is 2.5 million light years off.

However, the actual size of the universe is a mystery. We can see 46 billion light years out into space, but, as we shall see, the universe probably extends far beyond that.

## Is the edge of the universe a red-light district?

You'll remember from when we looked at waves that, just after a teenager in a boombox on wheels passes you, you hear the pitch of that bass he's pumping out become even lower because of the Doppler effect (see page 95). As relativity predicts, the same happens with the light from stars that are moving quickly away from us: the light waves from them seem to us to have a lower frequency—and therefore appear redder—than they would have been for someone not moving relative to the star. (Physicists say the light from retreating objects is *red-shifted*.)

It turns out that the farther out into space we look, the faster the galaxies are moving away from us, on out to the farthest galaxies we can see, which are now that as-far-as-we-can-see 46 billion light years away. The red shift of the light from these most distant galaxies tells us that they're moving away so quickly that the light we now see left them when they were just 36 *million* light years away from us. (Beyond these galaxies, we calculate that there are galaxies moving away from us so fast that the light from them can't ever reach us.)

This all means that everything in the universe is not just getting farther away from us: It's getting farther away from everything else, too. So the whole universe is expanding, which means it must once have been smaller. If we run time backward, we come up with a date, 13.7 billion years ago, when the entire universe was very small and very dense. (The Earth, by contrast, is just 4.5 billion years old.) So we need some event to create that small dense universe and turn it into a rapidly expanding one.

That event was christened the *Big Bang* in 1949 by the British astronomer Fred Hoyle. (Hoyle actually used "Big Bang" as a snappy way to refer to a theory he thought was totally wrong. Unfortunately for Hoyle, it was his alternative theory that was wrong.) But it wasn't until 1967 that proof of the Big Bang was discovered: It's called *the cosmic microwave background* (or CMB for short).

## Why do I feel this warm glow?

If it isn't caused by your enjoyment of physics, it may be due to the Big Bang.

As we saw when we looked at matter, something squashed and dense (like the universe as it was being born) must also be very hot. And hot things give off energy as electromagnetic radiation. And hot things as hot and big as the Big Bang give off a great deal of that radiation. In fact, the Big Bang gave off so much energy that, although it's now quite weak, it is still all around us—as the cosmic microwave background, a slight warm glow that fills all of space (if you look using a radio telescope).

We know the CMB is the afterglow of creation because it's special: It's the same in all directions, and its spectrum, with microwaves as its most common component and little visible light, is that of radiation from a black body (see page 140). Nothing but a hot, dense, uniform, exploding plasma-thing-that-will-be-a-galaxy could have made the CMB.

So, the universe began with a bang. Then it became a rapidly expanding bundle of dense, hot, opaque gas-like plasma, until, after about 380,000 years it had cooled enough for atoms to form. At that point, it became transparent, so that all the heat radiation flying around could

keep flying around. And that's what our CMB is: a post–Big Bang glow.

The CMB is almost the same in all directions, but there are tiny differences in temperature here and there, which correspond to tiny differences in density. Those differences were vital, because as the universe expanded and cooled, the slightly denser areas had greater gravity, so they pulled in more gas and became denser still, and slowly the galaxies and the stars within were formed.

In the stars, 10 billion years of nuclear fusion then turned the hydrogen and helium of the early universe into other elements, until they used up the hydrogen. At this point, most of these poor dying stars would have swelled up into what we call red giants (even though they're usually orange). Then, after just a few million years, the red giants would have thrown their outer shells off into the galaxy, spreading around all kinds of useful stuff that gathered to build new, more complicated stars, as well as planets packed with useful stuff like oxygen, carbon, iron, and uranium.

## Can we escape?

As we look out into the universe—and we can now see evidence for about 125 billion galaxies—it looks the same in all directions. There is no edge and no center, and the uniformity of the CMB also suggests that there is no center. So although the term "Big Bang" is snappy, it does give us the wrong idea. In a way, the Big Bang wasn't an explosion *in* space, it was the explosion—and the creation—*of* space.

Today it seems likely that wherever you sit in the galaxy, you will see all the universe expanding away from you. Yet this is not just because all the other galaxies are traveling through space. Instead it is because, as Einstein discovered, space-time itself can change shape, and the expansion of the universe is due to the expansion of space itself.

This is why, when you add up lots of expanding space between us and them, some galaxies can be moving away from us at speeds faster than light without breaking relativity. This is also why, although the universe is probably not infinite, it is unlikely (even if you could go faster than light) that you could ever find the edge. Instead, you'd curve around and (eventually) find yourself back where you started.

## What caused the Big Bang?

So far, we have direct evidence for what the universe was like after 380,000 years. Is that as good as physics can do?

Actually, with theory and with experiments in particle accelerators that mimic tiny bits of the early universe, we think we can look back much further. A second after the Big Bang, protons and electrons, and the other building blocks of atoms, had begun to appear in an 18,000,000,000°F (10,000,000,000°C) super-hot soup of subatomic particles. Further back still, after a tiny fraction of a second, the forces and laws that hold the universe together (including gravity and electromagnetism) had appeared. Back then, our soupy universe was hotter still, though it was starting to cool as some of the huge energy from the Big Bang began to be frozen, by the power of $E=mc^2$, into the first raw building blocks of matter.

And before the soup course there was just the oven: the Big Bang itself, a huge flash of energy that may have been caused by a strange kind of field (that's a field like a magnetic field, not one with grass and a cow it) called the *inflaton* field.*

This inflaton field, which basically came out of nothing, very quickly swelled the new universe from a tiny size (where the random effects of quantum mechanics were important) to one where the randomness was so spread out as to be almost invisible—except as those tiny but important variations in the CMB that tell us how the stars were formed.

Sounds crazy, doesn't it? But behind this story is a powerful theory, which was invented by the American theoretical physicist Alan Guth in 1979. And the math of the story—and the way it predicts exactly the minute variations in the CMB that we can measure—means that it holds up very well. Today, among physicists, inflation theory is pretty well tested and accepted.

## Is this the only universe?

Inflation theory says that the portion of the universe that is observable to us today expanded from a size of $10^{-50}$ meters (that's 0.00...49-zeroes-here-please...001 meters, which is way, way smaller than an atom) in radius at $10^{-35}$ seconds after the birth of the universe to almost one whole meter in radius at $10^{-34}$ seconds. (Just think for a moment

---

*That's infla-*ton*, like infla-*tion*, which is what the field caused, but spelled wrong so that it matches the spelling of other similar fields, such as the pro-*ton* field.

about how incredibly fast that expansion of space is. It is, for example, much faster than the speed of light. In fact, each time an inflating universe doubles in size, which happens many times, light doesn't even get time to travel from one side of an atom to the other.)

And Guth's theory is even crazier than it seems at first. Guth believes that inflation didn't just happen once. Instead, it is happening continuously, with new universes being born like bubbles in their own tiny nicks of inflationary time, then cooling as their bit of the inflaton field relaxes. And as a universe cools, it expands more gently, inventing its own laws and its own ways of doing things at its own gentle pace—and it is then cut off forever because space outside it keeps on madly inflating, apart from where it forms countless other tiny bubble universes, far, far away.

Guth even believes you could make a new universe in the lab, from a few grams of matter and a special kind of emptiness called a *false vacuum*. Well, maybe you could, but you definitely shouldn't. The huge energy released into our universe would be like 30 of the nuclear bombs dropped on Nagasaki all going off at once. And that's going to hurt, even if you've got your plastic lab glasses and rubber gloves on.

## Was it a bug-eyed monster?

The problem with inflation theory is that it is impossible to test. Whichever Big Bang theory you choose (and there are others), they agree that you can't see beyond the Big Bang, or out of our bubble universe. In fact, some would say, it makes little sense to think of what created our universe. The creation is more like an impassable boundary to our universe than an act in it.

But if you're bored with physics, you could come up with your own story. Maybe our universe was created by some bug-eyed monster who'd discovered the same laws as Alan Guth, a slimy bug-eyed monster who just happened to have some false vacuum handy. But remember, if it was, that alien got blown up a long, long time ago, and anyway, there is no possible way for us to know anything about its universe, which may have been very different from ours.

Actually, if Guth is right, and the creation of the universe is basically a free lunch, it seems likely that the vast majority of universes are probably too weird to support anything like life as we understand it. That's because there's only a tiny chance that any universe that coalesces out of the hot, dense post–Big Bang soup will be as cozy for life as ours. So although our universe is vast, violent, and unforgiving, we can also see that if gravity was only slightly less strong, or if there were 5 or 7 or 27 dimensions, or if *dark matter* was less common, then balloons might fall and apples might rise—or, more likely, galaxies and stars could not have formed and there would be no apples, no balloons, and no humans to miss them.

## But what is dark matter?

Well, for a start, dark matter isn't dark. It is in fact completely transparent. It's just that space is dark, so we can't see it.

When we look out at the universe and add up all the mass we find, there just isn't enough to make enough gravity to help galaxies form and move like they do. In fact, we now estimate that we can see only 5 percent of the stuff in the universe. The rest is dark.

There must, therefore, be huge clouds of invisible stuff, stuff that light and other forms of electromagnetic radiation pass straight through because it's not made up of protons and neutrons at all, but that supplies the missing gravity to give galaxies the right structure.

It's not just theory. We've seen light from distant galaxies bent by what must be dark matter. But what exactly dark matter is, and how it came to be all over the place, is a mystery. Physicists can't even agree on whether it's cold, or warm, or hot. Worse still, much of the dark stuff in the universe is probably not dark matter: instead, it's dark energy, spread throughout space and, according to general relativity, forcing the expansion of space to happen more quickly than it might. We can measure this increase in the expansion of space. But we don't have a clue about what dark energy is. We know it's dark and we know it's energy and er...that's it.

This is one place where physics is still in the dark ages. Our current best model of the universe—the Standard Model—sheds no light on dark matter, dark energy, and a few other important things, such as gravity. So although we have plenty of theories about why balloons rise and apples fall, at their most basic level the two theories are separate. Apples fall because of gravity, which is explained by relativity. But balloons rise because of temperature and pressure, and that ultimately is caused by quantum physics and the Standard Model.

# Why are so many physicists growing a GUT?

Even slim, fit physicists may be working on a Grand Unified Theory (GUT) or a Theory of Everything (TOE): one that will add in what the Standard Model leaves out.

The Standard Model of physics uses the interaction of tiny particles to explain matter, electromagnetism, the strong and weak nuclear forces inside atoms, and how those things affect each other. Remember all those forces we spent so long discussing? Well, there's no such thing—at least on the quantum scale.

First, there are the 12 *fermions*—the particles that make up matter. What happens when these particles meet and interact with light, and how they join together to form atoms, is then explained by 12 force-carrying particles called *bosons,* which fly around at the speed of light, between the fermions, pushing and pulling them around. The bosons include the photon, the particle that carries what we observe as a field of force between things that have electric charge. There's also an elusive 13th boson, the Higgs boson, which gives the other elementary particles their mass.

The problem is that once you try and add in what the Standard Model leaves out, and formulate a GUT, everything gets much more complicated. Some progress has been made, but as yet there is no evidence at all that gravity depends on a new kind of fundamental particle called a graviton—and it seems impossible to find one, other than in a huge particle accelerator such as the Large Hadron Collider built by the European Orginization for Nuclear Research (CERN).

One avenue of research depends on *supersymmetry*—which involves adding in a gaggle of extra particles that no one has ever seen and which creates a whole new set of problems that are possibly answered only by yet another theory, called *string theory*. But it relies on notions like 11-dimensional space and a lot of math that is so complicated and so new that even most physicists can't understand it. And there is no way to test string theory because the effects that it predicts take place at very high energies and over very tiny distances.

There are alternatives to GUTs and TOEs and getting tied up in strings. One was created in the 1990s by Mark Hadley, at Warwick University in England. It's lean and mean and uses Einstein's space-time to explain quantum physics. No one can find much wrong with it, but it's pretty much ignored. The reason? Well, it could be argued that a thousand scientists are working on their GUTs simply because a thousand scientists are working on their GUTs—and because the best way to get funding for your GUT work is for a bunch of other scientists to agree that what you're working on is worth the massive grant your GUT needs.

This doesn't mean that our best GUTs and TOEs are wrong. Maybe we will find ways of checking them indirectly. Or maybe, as the next question shows, our best physics isn't quite as good as we'd like to think.

## Can we travel faster than light?

No. Of course not. At least, that's what Einstein argued. So it seems we are stuck in our solar system, forever forbidden by the laws of physics from building a Starship *Enterprise* and seeking out new life and new civilizations.

But actually, in quantum physics there does seem to be something that happens faster than light: communication between what are called *entangled particles*. These entangled particles are pairs of particles emitted when one particle decays (either naturally, or because a physicist smashes it) into two new particles, so that if one has spin of up, the other has spin of down.* But quantum physics also says those particles have no definite spin at all until that spin is measured. And yet, once you do the experiment, and Alice measures the spin of one particle, Bob will find that the spin of the other is opposite, even if Alice and Bob are a long way from each other.

So it seems that, once the spin of one particle is measured, it immediately communicates to its twin what its spin should be. (And when we say immediate here, we mean immediate. Experiments have shown that the communication between entangled particles happens at speeds millions of times faster than the speed of light.)

Physicists tend to argue that actually this experiment does not break special relativity, because nothing in it really moves faster than light. They say that twin particles could not, for example, even be used to communicate information, let alone help something as big as us zip to the other side of the universe through a *Star Trek*–style teleporter.

But that doesn't stop entanglement from being a real problem for current physics. And it doesn't necessarily stop us boldly going and meeting Vulcans, either. Because we still we have one last, best hope for intergalactic travel: a science-fiction shortcut called a wormhole.

---

*Spin is a strange property of tiny quantum particles. It's complicated and doesn't involve anything actually spinning around.

## Can we jump through a wormhole?

Wormholes are space-time tunnels that could, in theory, lead directly to far-off galaxies (or far-off times). By traveling through a wormhole, you could go as far as you want and arrive even before you left. And the good news for budding Stargate SG-1 officers is that the laws of physics, as we understand them, say wormholes aren't pure sci-fi: They could exist. The bad news, though, is that the laws of physics also say that wormholes must be very unstable, and as soon as you try to stuff anything as big you, or your spaceship, or even an atom, through one, it will collapse.

Which leads us to the biggest physicist-stumper of them all...

## What is a law of physics?

The idea that there might be laws of nature originally came from thinking about the laws that societies create. The laws of physics were seen as a vast force of cosmic policemen, keeping order among matter and everything else in the universe.

Christianity fired the police force, because it had one powerful God to do all the law making and keeping, and because the universe seemed so well behaved. Newton thought that the simple, elegant, mathematical laws he discovered were thoughts in the mind of a straight-thinking God, and that they were nice proof of just how orderly and rational He was.

Today, the fact that the fundamental laws are small in number (even if the quantum ones are pretty weird) is just part of why few physicists still think we need a god to keep things in order. Instead, the laws have come to be seen as fundamental properties that create the entire universe (and us in it), just like a few simple rules mean water molecules can freeze together in an infinite and beautiful variety of snowflakes.

But the importance of math in physics makes it easy to dodge the question of what the laws of physics are. Lots of physicists like the slogan: "Shut up and calculate!" They plug the numbers in: the acceleration due to gravity; the number of apples in the basket; the size of the balloon; the temperature of the air inside. They make the prediction: balloon to rise to 1.5 miles (2,500 m). They wear that extra sweater. And if they're warm and toasty when they go on their balloon trip, then they like the law.

But the laws of physics aren't only about getting your calculations right. The best laws also explain why the calculations work. For example, in 1915 the German mathematician and physicist Emmy Noether discovered how the conservation laws—which, as we've seen, are some of the most basic and important in all of physics—arise directly from the structure of time and space.

For example, the fact that the laws of physics are the same *at all times* gives us the law of conservation of energy. And the fact that the laws of physics are the same *everywhere* means that momentum must be conserved. And once you know this, and once you understand the shape of space-time, as Einstein did, special and general relativity fall into place. All it takes is knowledge, creativity, and maybe a touch of genius. It's amazing what the human mind can do. And it's even more amazing that we've tested and tested these ideas and they seem exactly right. So maybe that gives us some sort of answer to the question about laws. But it doesn't end the debate.

Noether proved her theorem using math. And theoretical physicists like Einstein routinely come up with new ideas by creating new math, long before these new ideas can be tested in the lab. So, even deeper in physics than laws that explain the laws is the math that explains those basic laws. And the reason why math matches the way the universe works is the deepest mystery of all.

## Are there any more questions?

Of course there are. Books full of them. If you ever get the chance to stump a real physicist, you could be talking for hours and hours...

Despite how far we've come, physicists aren't running out of things to do. They are still driven by the lure of discovery: the lure of uncovering a little more of the beauty and strangeness of the world, and of being, for a moment, the first and only person ever to know some new thing, and at the same time, knowing that they alone have found that new thing—and that they have found it purely through the power of their own thought.

Isn't physics fun?

# INDEX

## A

absolute zero, 86
acceleration and deceleration, 26, 35–40
air, 32, 82–83
aircraft, 30
alternating current (AC), 112, 115–17. *See also* electric current
angular momentum, 68–69, 71
anti-matter, 153
apples, 25, 44, 46–47
Archimedes, 21, 31–33
Aristotle, 9, 10, 16
astronomy, 11, 12
atmosphere of Earth, 82–83
atoms, 13, 15, 19, 26, 66, 76, 77, 79, 137, 145–48, 160

## B

balloons, 31–34
Balmer, Johann, 145
batteries, 107–9, 110
bicycles, 68–69, 70–71
Big Bang, 7, 134, 158, 160–61
black bodies, 140, 145
black holes, 131–32, 134–45
boats, 30–31
Bohr, Nils, 145, 146–48
Bose-Einstein condensates, 79, 86
bosons, 165
braking distances, 48–56
Brown, Robert, 81
Brownian motion, 81–82, 139
buoyancy, 30–34

## C

calories, 42–43. *See also* Joules
car crashes, 48–56
centripetal force, 34–35
chemical energy, 42
chemistry, 7, 76–77
color, 98, 101, 103, 144–45, 148–49
conductors and conduction, 107, 109, 112–14, 115
conservation laws, 62, 169
conservation of energy, 19, 44, 47, 62–63, 95, 169
convection, 85
cosmic microwave background (CMB), 158–59, 161
cosmology, 14, 134. *See also* Big Bang
*Crouching Tiger, Hidden Dragon,* 57–58, 59, 72–73, 74
curvature of time and space, 131–35

## D

Dalton, John, 76
dark matter, 163–64
Davy, Sir Humphrey, 108
density, 32–33
Descartes, René, 100
diffraction of waves, 93–94
Dirac, Paul, 149, 153
direct current (DC), 111, 117. *See also* electric current
direction of forces, 34, 39, 68
Doppler effect, 95–97

## E

Earth, 13, 39–40, 157. *See also* planets

Edison, Thomas, 108–9, 114–15
Einstein, Albert, 14, 119–36,
    138–39, 143
electric charge, 63, 81, 97, 105–8,
    110, 111, 113, 165
electric circuit, 108, 110–11
electric current, 109–17
electric fields, 97
electricity, 91, 105–17
electric potential, 110
electric power, 112, 114, 115, 117
electrodes, 108
electromagnetism, 14–15, 98–99,
    112–14, 140, 145, 158
electrons, 13, 105–9, 145, 147–48,
    160
electrostatic discharge, 107
elements, 77, 144, 148
E=mc², 127, 139
emission spectrum, 144–45, 148
energy, 41–56. *See also specific kinds*
    conservation of, 19, 44, 47,
        62–63, 95, 169
    gravity and, 132
    mass and, 14, 127–29, 139
entangled particles, 167
ether, 9, 122, 123
exact laws, 62

**F**
falling, 24–25, 29–30, 129–30
Faraday, Michael, 113–14
feathers, 24, 29, 30
Fermionic condensates, 79
fermions, 165
floating. *See* falling
forces, 23–40
Franklin, Benjamin, 106, 107
frequency, 96, 143. *See also* waves
friction, 39, 49–51. *See also* resistance

**G**
Galileo, 12, 16, 120–21
gases, 78, 79, 148–49
glaciers, 58–59
golf balls, 39

*Good, the Bad and the Ugly, The,* 67
GPS satellites, 134
Gramme, Zénobe, 114
Grand Unifed Theory (GUT),
    15–16, 165–66
gravity, 12, 16, 25–26, 35, 132
    angular momentum and, 68–69
    relativity and, 129–30, 133–34
    time and, 133–34
Greeks, ancient, 9–11, 106
guns, 67
Guth, Alan, 161–62
gyroscopes, 70

**H**
Hadley, Mark, 166
hammers, 29–30
heat, 14–15, 42, 75–87
    radiation of, 14–15
Heisenberg, Werner, 149
Hertz (Hz), 96
Higgs boson, 165
history of physics, 7–21
Hooke, Robert, 8
Hooke's Law, 8, 17–18, 19
Hoyle, Fred, 158
Hubble, Edwin, 156
Huygens, Christian, 100

**I**
inertia, 73–74
inflation theory, 161–62
inflaton field, 161
interference
    light and, 100–101, 103
    of waves, 91
invisibility, 102

**J**
Joule, James, 43
Joules, 42–43

**K**
kinetic energy, 42, 49, 49–51, 117,
    127

**L**

laws of physics, 8, 168–70. *See also specific ones*
light, 97–99
    black holes and, 135
    diffraction and, 100–101
    electric, 109
    frequency, gases and, 148–49
    gravity and, 133
    interference and, 100–101
    particles, 15
    red-shifting and, 157
    speed of, 36, 97, 122, 127
    as wave and particle, 99–104, 140, 150–51
lightening rod, 107
liquids, 79

**M**

Mach, Ernst, 138
magnetism, 97, 113. *See also* electromagnetism
magnitude, 35
mass, 32–33, 37–38, 60, 73–74, 132
    energy and, 14, 127–29, 139
mathematics, 11, 38, 51, 96, 161, 170
matter, 75–87
    dark, 163–64
    fermions of, 165
Maxwell, James Clark, 98
Maxwell's equations, 98
meter, length of, 97–98
metrics, conversion of, 175
Michelson, Albert, 123
Millikan, Robert, 143
momentum, 57–74, 95, 132, 169
Morley, Edward, 123
motion, Newton's laws of, 12–13, 27, 37–40, 61

**N**

neutrons, 13, 147
Newton, Sir Isaac, 12, 16, 100, 104, 169
    laws of, 27, 37, 38

vs. Einstein, 135–36
Noether, Emmy, 169–70
nuclear bombs, 128–29
nucleus of an atom, 147–48

**O**

Ohm, Georg, 110–11
Ohm's Law, 111
Ørsted, Hans Christian, 112

**P**

particle accelerators, 160
Pascal, Blaise, 84
Pascals, 83
photons, 104, 125, 142–43, 147, 150–51
photosynthesis, 42
physical constant, 97
Planck, Max, 14–15, 141
Planck's constant, 141
planets, 11, 133. *See also* Earth
plasma, 79
pool balls, 60–65, 66, 72–73
potential energy, 47
power, 41–56
precession, 71
pressure, 82–83
protons, 13, 147, 161
Ptolemy, 11
puddles, evaporation of, 86–87
Purley, David, 55

**Q**

quantum, defined, 142
quantum physics, 15, 16–17, 18, 137–53
quantum tunneling, 153

**R**

radioactivity, 128–29, 147
rainbows, 101
rarefaction, 95
reflection, 93
refraction, 92, 101
relativity, 14, 16–17, 119–36, 139, 169

resistance, 110–11
air/wind, 27–28, 29–30, 39, 47
   electrical, 110–11
Richmann, George, 106
Rutherford, Ernest, 145–47

**S**

scalar quantity, 28
Schrodinger, Erwin, 149, 152
scientific revolution, 12–13
Scott, Dave, 24–25
sea power, 91
solar energy, 41–42
solids, 79, 80–81
sound energy, 36, 42, 49–50, 95–97
space, curvature of, 130–31, 132
special relativity, theory of, 14,
   120–26
speed. *See* velocity
spin, 167
springs, 17–18, 87
Stapp, John, 53, 56
Stapp's law, 53
stars, 11, 14, 32, 159
static electricity, 106, 107, 109
static friction, 51
stopping distances, 48–56
string theory, 166
sun, 32, 41–42
supersymmetry, 165–66

**T**

terminal velocity, 30
Tesla, Nikola, 21, 112, 114
Theory of Everything (TOE), 165–66

thermal energy. *See* heat
thermodynamics, 44, 86
Thomson, J. J., 107–8, 146
time, 125–26, 130–31, 133–34
torque, 70, 71
transformers, 115, 116
transverse waves, 94
travel, speed of light and, 166–67

**U**

uncertainty principle, 149
universe, 14, 155–70

**V**

vector quantity, 28
velocity, 30, 34–35, 60, 127–28
Volta, Alessandro, 110
voltage, 110
volume, 32–33, 79–81

**W**

water, 32, 77–81
watts, 46
Watts, James, 46
wavelength, 94, 95
waves, 14–15, 89–104
weather, 42
weight of an object, potential energy
   and, 47
work, defined, 45
wormholes, 168

**Y**

Young, Thomas, 100–101

# METRIC CONVERSION TABLE

| TO CHANGE: | INTO: | MULTIPLY BY: |
| --- | --- | --- |
| Millimeters | Inches | 0.039 |
| Centimeters | Inches | 0.394 |
| Meters | Feet | 3.28 |
| Meters | Yards | 1.09 |
| Meters per second (m/s) | Miles per hour (m.p.h.) | 2.25 |
| Kilometers | Miles | 0.621 |
| Square centimeters | Square inches | 0.155 |
| Square meters | Square feet | 10.8 |
| Square meters | Square yards | 1.2 |
| Cubic centimeters | Cubic inches | 0.061 |
| Cubic meters | Cubic feet | 35.3 |
| Cubic meters | Cubic yards | 1.31 |
| Liters | Pints | 2.11 |
| Liters | Quarts | 1.06 |
| Liters | Gallons | 0.264 |
| Grams | Ounces | 0.035 |
| Kilograms | Pounds | 2.2 |
| Metric tons | Tons | 1.1 |

entrifugal Force • Electr

gy • Inertia • Quantum M

Newton's Laws • Ohm's I

adioactivity • Levers • Th

ynamics • Acceleration •

elocity • Vector • Gravity

mplitude • Alternating Cu

Reflection • Atom • Ato

athode Rays • Centrifuga

onvection • Decibel Scale

agnetic Force • Electro

amma Ray • Horsepower